MW01629110

Modern Chlor-Alkali Technology
Volume 7

Modern Chlor-Alkali Technology
Volume 7

Edited by

S. Sealey
Associated Octel Co. Ltd., Ellesmere Port, UK

The proceedings of 1997 London International Chlorine Symposium organised by the SCI Electrochemical Technology Group, held on 4–6 June 1997, at The Gloucester Hotel, London.

Special Publication No. 221

ISBN 0-85404-723-9

A catalogue record for this book is available from the British Library

Published for SCI by The Royal Society of Chemistry,
Thomas Graham House, Science Park, Milton Road,
Cambridge CB4 4WF, UK

For further information visit the RSC web site at www.rsc.org

Printed by Bookcraft (Bath) Ltd

Introduction

The papers in this book were presented at the 1997 London International Chlorine symposium. This was the eighth symposium organised by the Electrochemical Technology Group of the Society of Chemical Industry, held at three year intervals.

The conference was attended by 204 delegates from all six continents and the popularity of the symposium remained high with 39 papers offered, 26 of which were accepted for inclusion in the programme.

The organising committee wished to consider developments in production technology. Topics covered include the latest developments in membrane and mercury technologies, together with improvements to diaphragm cells. The fundamentals of chlorine evolution reactions applicable to all technologies are discussed, along the removal of brine impurities and aspects on improving the control of processes and operator training.

We hope that this book will provide an exciting overview of the latest technologies in use within the chlorine industry as well as emphasise the vital issues of public awareness, environmental aspects and the need for good neighbour relations.

The Electrochemical Technology Group of the SCI would like to thank all delegates who attended the conference, and whose participation in discussion was important to its success. In particular, it would like to thank the following.

1. The contributors of the papers.

2. The session Chairmen.

   Mr M E Lyden (The Chlorine Institute)
   Dr M Cooper (ICI Chemicals and Polymers Ltd)
   Mr S A Sealey (The Associated Octel Company Ltd)
   Mr J S Moorhouse (Rhone-Poulenc Ltd)
   Mr T Wellington (Hays Chemical Distribution Ltd)

3. The Chlorine Institute and Eurochlor for assistance with printing costs.

4. The Associated Octel Company Ltd., for secretarial support and hospitality.

5. The Organising Committee.

   Dr A T Clarke (Hays Chemicals Distribution Ltd)
   Mr J S Moorhouse (Rhone-Poulenc Ltd)
   Dr R Pilkington (ICI Chemicals and Polymers Ltd)

S A Sealey
Chairman - Organising Committee
The Associated Octel Company Ltd

Contents

1

CHLOR-ALKALI: IS THE CYCLE ALIVE AND WELL?

M. G. Beal

Harriman Chemsult Limited
45 Britton Street
London EC1M 5NA, UK

1 INFORMATION SOURCES

Harriman Chemsult is a chemical industry consulting company whose roots are in the chlor-alkali sector. Our activities fall into four major areas:

Price monitoring of chemicals	Weekly basis
Detailed reports on chlor-alkali	Monthly basis
Single-client reports	On request
Multi-client reports	Intermittent

In addition to chlorine and caustic soda, our activities encompass related chemicals such as soda ash, PVC, VCM, EDC, sodium chlorate, hydrogen peroxide, sodium hydrosulphite, etc. The daily contact established with the industry, and frequent personal meetings with chemical producers and customers has enabled us to build an enviable in-house database of knowledge on the chlor-alkali and related industries. It is from this in-house store that all the data in this chapter has originated.-

2 THE CHLOR-ALKALI CYCLE

It seems sensible to first of all explain the title of this chapter before moving on to discuss in detail the historical perspective and offer some pointers to the future.

Chlorine and caustic soda are produced in almost equal volumes via the electrolytic decomposition of sodium chloride (common salt). There are a number of other processes by which either of the two materials can be produced singly, but the electrolytic decomposition of salt accounts for probably 98% of all chlorine and caustic soda tonnage produced globally. This situation seems unlikely to change. While chlorine and caustic soda are produced in almost equal quantities globally, the end use markets for each product do not coincide. Figures 1 and 2 below illustrate end use sectors for chlorine and caustic soda in the USA for 1995. The breakdown of demand by end use for Western Europe is not significantly different. Since these two regions of the world account for around 60% of global chlor-alkali demand, the USA end use breakdown can be taken to be representative of global demand for the products. The point of presenting the two charts is not to concentrate on the fine detail, but to illustrate the very different industrial end uses of chlorine and caustic soda.

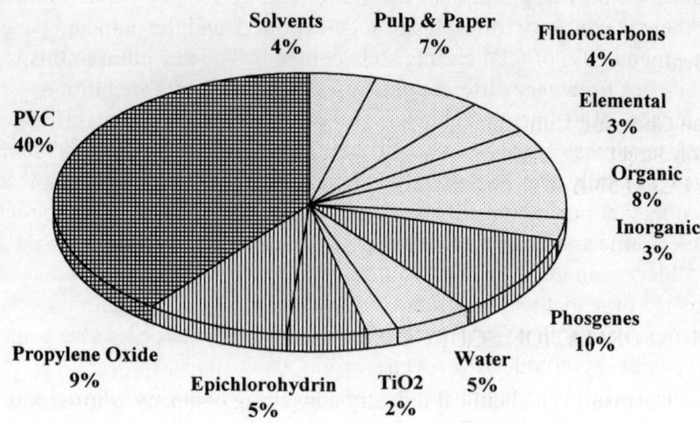

Figure 1 *USA Chlorine End Uses, 1995 (Total = 10.192 million MT)*

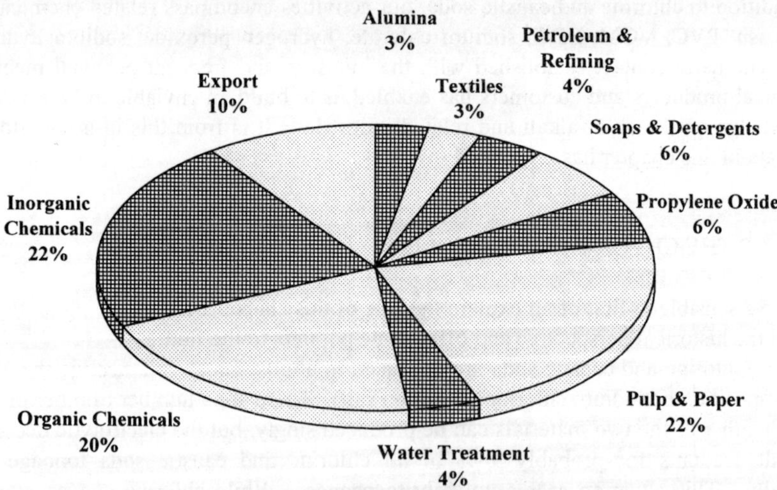

Figure 2 *USA Caustic Soda End Uses, 1995 (Total = 11.525 million MT)*

It should be noted from the chlorine end use chart (Figure 1) that plastics industry end use accounts for 64% of chlorine demand. This encompasses plastic where the chlorine ends up in the product (the PVC chain), and plastic where the chlorine does not remain in the finished product, i.e. propylene oxide, epichlorohydrin and phosgene end uses. Most of the demand for the latter three materials occurs within the plastics industry. For interest, we estimate that the plastics industry accounts for around 60% of chlorine demand in Western Europe.

It can be seen from Figure 2 that 42% of caustic soda demand originates in the chemical industry, in a very diverse set of end uses. Otherwise, the pulp and paper sector and exports are the most important sectors for caustic soda demand. For interest, in Western Europe there is now very little caustic soda exported, and the pulp and paper sector accounts for only around 10% of total caustic soda demand. The key point to this is that caustic soda demand comes from very different industries than does that for chlorine.

In view of the fact that chlorine and caustic soda have very different end uses, but are produced in almost equal volume, it can be seen very easily that demand for the two products will hardly ever be perfectly in balance, and that the demand imbalance arising will influence market perceptions of the value of the two products. It is the demand imbalance between chlorine and caustic soda that is at the root of the cyclicality to be discussed.

Chlorine and caustic soda can be regarded as a principal product and a by-product. Which is the principal and which the by-product changes through time. As soon as demand for chlorine exceeds that for caustic soda, chlorine is regarded as the principal product and caustic soda the by-product; the reverse of this also applies.

The major manifestation of the imbalanced demand for chlorine and caustic soda is a highly volatile spot price, particularly for caustic soda. We intend to discuss this phenomenon in some detail. We observe the most volatile movements most easily in caustic soda prices because greater volumes are traded than of chlorine. Spot prices, while covering a smaller section of the market than contract prices, give an early indication of demand imbalances, and do influence subsequent negotiations of contract prices.

Over the past 25 years we have observed a number of caustic soda price cycles. These appear to have a periodicity of 6-7 years, and are illustrated in Figure 3.

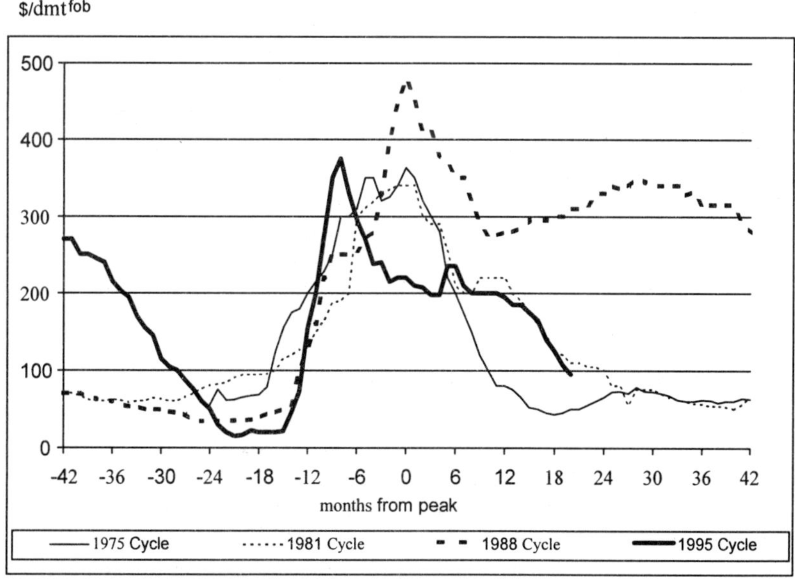

Figure 3 *USA Caustic Soda Liquid, Export Price Cycles Superimposed*

While it is obvious that there is no mathematical equality in each of the price cycles observed, that some kind of relationship exists is indisputable. The last price cycle that peaked in 1988 differed from those of 1975 and 1981 by having an extended period of relatively high caustic soda prices long after the price had fallen from its peak. In the current (1995) price cycle, caustic prices have fallen a long way off the peak, but as this report is being written in mid-1997, caustic soda prices appear to be edging upwards. The important question is whether the current uptick will lead to a prolonged period of moderately high prices, followed by a complete slump in 18 months time, or have we seen a shortening of the periodicity of the cycle, and are caustic soda prices about to surge upwards towards a new peak? It is simply too early to tell at present, and the non-mathematical nature of the cycles observed means we cannot make definitive predictions.

What is interesting is an examination of previous predictions in the chlor-alkali industry, and reflection on what we might learn from these predictions today.

3 CHLOR-ALKALI CYCLE DEVELOPMENTS 1983-1997

To conclude our presentation at the 1994 London International Chlorine Symposium[1] we suggested there were really only four possible forecasts that could be made. Any other scenario would simply be a variant of one of these four. The four scenarios are:

i The future might be very similar to the past, with alternating periods of chlorine tightness and surplus, resulting in large fluctuations in pricing for both caustic soda and chlorine.

ii Demand for PVC and chlorinated derivatives might remain strong, whilst caustic soda demand advances only slowly, leading to a continuing caustic soda surplus, and thus low prices for caustic soda.

iii Demand for chlorine and its derivatives might embark on a lengthy or even permanent downhill path. This would be encouraged by negative publicity expounded by the environmental lobby. Under this scenario, caustic soda supply would be tight, prices would increase dramatically, and expansions in capacity for soda ash causticisation would be planned.

iv Demand for chlorine and caustic soda might enter a period where demand for both products would remain approximately balanced, leading to prices for the two chemicals moving in tandem rather than in opposing cycles.

We favoured the first scenario as a forecast of future behaviour of chlor-alkali. However, at various times in the past it has been fashionable to four one of the other scenarios, in particular scenarios (ii) or (iii). To understand why the industry has from time to time believed these other outcomes possible, we must look a little closer at the price development of chlorine and caustic soda. These are presented in Figures 4 and 5.

3.1 Chlor-Alkali Cycle Developments 1983-1991

It is interesting to reflect on the industry outlook six years ago, at the time of the 1991 London International Chlorine Symposium.

Chlorine spot prices had been on a downhill trend for three years (Figure 4), and the product was almost being given away. The background to this was both environmental and technology driven. Mercury and asbestos were both regarded as unsuitable materials to use for construction of chlor-alkali plants. The North Sea Conference had reached agreement

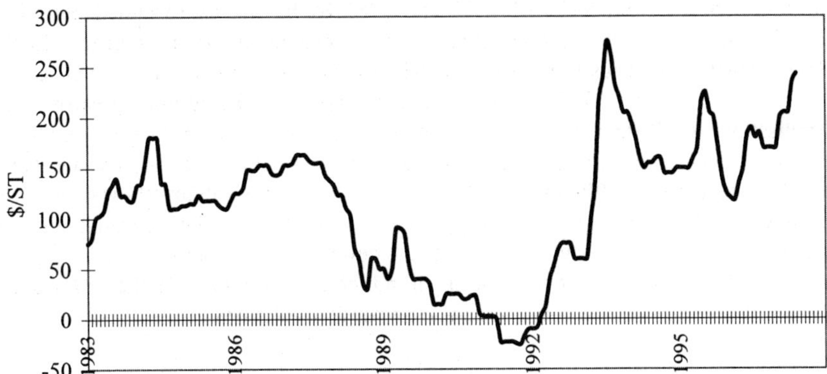

Figure 4 *USA Spot Price, Chlorine, 1983-1997*

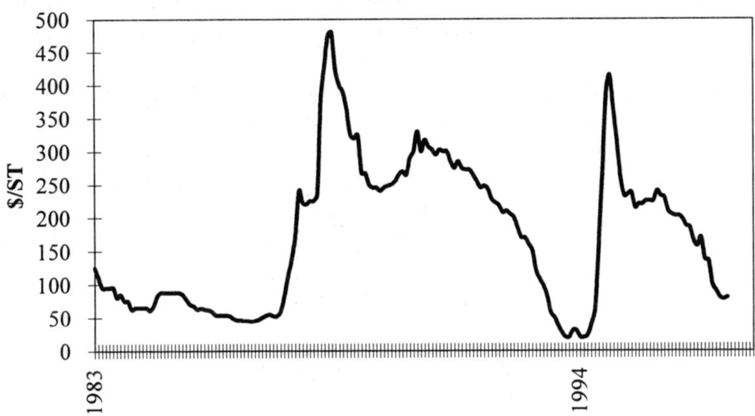

Figure 5 *USA Spot Price, Caustic Soda, 1983-1997*

that mercury cell plants would be phased out in signatory countries, and replaced by membrane cells. Chlorine had suffered a steep decline in demand in Western Europe for pulp bleaching, leading to the closure of some chlor-alkali plants whose output was almost dedicated to that industrial sector. New investment in chlor-alkali plants was viewed with extreme caution because of the perceived problem of how a use could be found for the chlorine produced.

In contrast (Figure 5), caustic soda had only three years earlier reached an all time peak, and although much reduced from the peak, in 1991 it seemed as if caustic soda might remain tight in the long term, that the cycle had been broken, and that prices would remain above $250/ST for a long time to come. Soda ash causticisation plants had been constructed in the USA on the back of this kind of forecast.

Overall, this 1991 outlook corresponds to scenario (iii) of our four potential forecasts. In time this kind of outlook proved to be incorrect on a long term basis, although it did describe the conditions that existed for a period of time.

3.2 Chlor-Alkali Cycle Developments 1991-1994

We should then reflect on the industry outlook three years ago, just before the 1994 London International Chlorine Symposium.

Figure 4 shows that chlorine pricing had recovered from its floor of 1991/92, and had peaked itself in 1993. The background to this now respectable pricing was the very strong demand for PVC, together with the fact that global chlor-alkali plant capacity had continued to decline since 1989. Some new capacity had been installed, but far more old capacity had been shut down, particularly in Europe. This is illustrated in the chart below (Figure 6).

On the other hand, as can be seen in Figure 5, caustic soda prices had been in a gradual decline since 1991, and had reached an all time low in the early part of 1994. Some would argue that contrary to the opinion expressed only three years before, the relatively weak market for caustic soda together with strong demand for chlorine into PVC may have swung the balance such that caustic soda was now destined to be the by-product on a long term basis.

At Harriman Chemsult we favoured the view that the chlor-alkali cycle was still functioning and that by 1995 caustic soda prices might well have recovered significantly. In fact this is precisely what happened, though at an even more rapid rate than we could have forecast. Within months of the 1994 conference, caustic soda spot prices had reached $400/ST, but as can be seen from Figure 5, the peak was fairly short lasted.

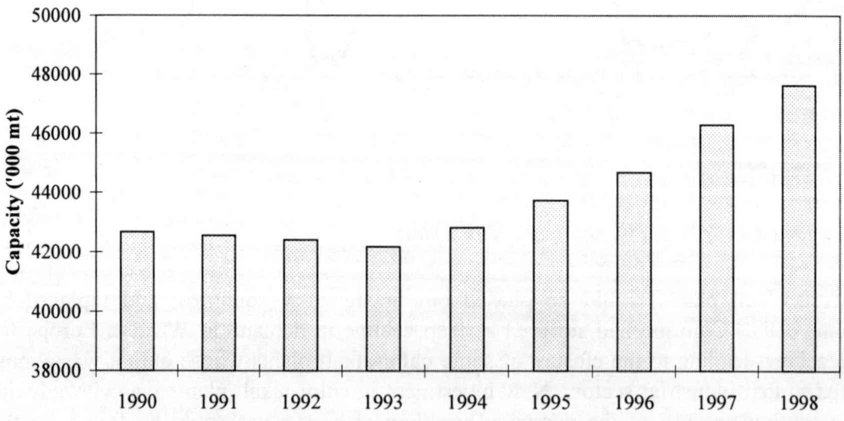

Figure 6 *Global Chlorine Capacity Development, 1990-1998*

3.3 Chlor-Alkali Cycle Developments 1994-1997 and Beyond

In order to formulate a view of where the market is currently headed, we need to look at the events since 1994. From Figure 4 it can be seen that chlorine prices while somewhat volatile, have held up rather well, with spot prices heading for a new peak in 1997. On the other hand caustic soda (Figure 5) appears to have passed through another cycle, with prices declining throughout 1996 and 1997.

As already discussed, the relatively strong chlorine market has been partly a reflection of the relatively low rate of increase in global capacity. However, since 1996 capacity appears to have been increasing once more, and we expect further capacity developments in the next few years.

It is interesting to note that in mid-1997 chlor-alkali producers are attempting to push through caustic soda price increases, while at the same time, chlorine is not showing any signs of weakness. In the short term we may see a period when the prices of both products increase at the same time. This rather unusual event might correspond with point (iv) of our potential scenarios. We do not believe that this scenario, if it occurs will be very long term. Eventually we believe increased chlor-alkali capacity will result in the resumption of the chlor-alkali pricing cycle seen in the past.

At this stage we need to comment on the spot prices used in this discussion. We are well aware that spot prices illustrate the extremes of pricing and do not reflect the prices at which most of chlor-alkali volume is sold to customers. However, spot prices do act as a very useful leading indicator of the direction in which prices are moving. In addition, contract negotiations will use spot price movements as part of the justification for setting new contract prices. In order to show that spot prices do have relevance to the discussion, we reproduce below a chart with USA chlorine and caustic soda contract prices superimposed.

The counter-cyclicality in pricing of the two products is less obvious, but can be seen in Figure 7.

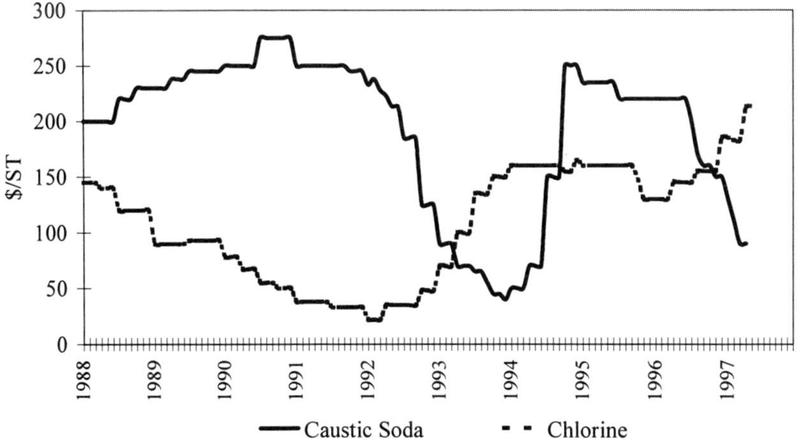

Figure 7 *USA Domestic Contract Prices, Chlorine and Caustic Soda, 1988-1997*

One interesting point is to speculate on where exactly we are in our suggested 6-7 year pricing cycle. If we re-examine Figure 3 it is clear that the 6-7 year periodicity could mean that the 1995 cycle has up to another 18 months to run yet. On the other hand, the caustic soda price chart indicates that this cycle is almost over. The question we need to ask is whether the cycle has speeded up and now has a periodicity shorter than 6-7 years, or is the current attempt to raise caustic soda prices merely a blip, and will prices remain low over the next year to 18 months?

4 CONCLUSIONS

In 1994 we were able to hazard the view that caustic soda prices were almost ready for a steep increase, and this is indeed what then happened. Today we have more difficulty in making a pronouncement, because there is conflicting evidence.

One aspect of which we are confident is that for as long as chlorine and caustic soda are produced by the electrolysis of sodium chloride, and arise in almost equal volumes, pricing of the two products is likely to exhibit some counter-cyclicality. This arises from the almost inevitable demand imbalance for chlorine and caustic soda. In reviewing the summary of conclusions to our 1994 paper, we still believe that scenario (i) remains the most compelling long term outlook for chlorine and caustic soda. We firmly believe that the chlor-alkali cycle is alive and well.

References

1. M. G. Beal, 'Modern Chlor-Alkali Technology, Volume 6', R. W. Curry (Ed), SCI, London, 1995, 1-12.

2
PUBLIC RELATIONS AND CHLORINE

J. P. Shields, J. A. Clarke and N. K. Sant

Hays Chemical Distribution Ltd
Sandbach
CW11 3PZ

1 BACKGROUND

The purpose of this volume is to report progress in Chlor-Alkali technology and we expect to hear of improvements and advances which will make the operation of our industry more efficient and safer for man and the environment. Similarly when the social, political and economic environment is discussed we should expect progress to be reported. The papers in this section develop the arguments of presentations on similar topics given at the last two meetings in 1991 and 1994.

Jim Shields presented a paper to the 1991 SCI meeting (Modern Chlor-Alkali Technology, Volume 5, p.35) concerning environmental aspects of chlorine bleach, sodium hypochlorite. He set out three premises on which he considered that the growth of environmental concern amongst the public was based. First was that public concern with the environment was not a matter of the welfare of animal or plant life but the belief that if those are threatened so too is personal well-being. A second premise was that the explosion in communications had created a demand for information, and third, that sensational equals newsworthy.

Six years later we can reflect that the cynical aspect of the first premise has been balanced by a genuine concern for the natural environment. The explosion of communications has continued unabated and we now have webs, networks, channels, news programmes which would not have been thought possible in 1991. While information is now more available, it is also more free of editorial or proprietorial influence. Chlorine producers can and do publish on the Net and their statements are as readily available as those of pressure groups, regulators or governments.

The third proposal, that to be newsworthy a topic must be sensational, still holds true for the chemical industry. We have hit the headlines during the intervening six years with recent events at Avonmouth, Knottingley, Castleford, Bradford and Ellesmere Port. These have not improved the image of an industry which is prevented from touching bottom spot in the popularity league only by nuclear power and reprocessing.

To review the top stories; Mad Cow Disease had hit number one in 1990, with a cover version hitting number one again in 1996. For the chemical industry, and chlorine in particular, the sensational aspects moved away from poison to endocrine modulation and stunted manhoods.

The ability of micro-organisms to get where they are not wanted, e.g. E.Coli, Cyclosporidium and super Staphylococci, has added to the newsprint use without praising the role of chlorine products in fighting them.

Scare stories are still a staple of the media, particularly when a threat to children or future generations can be suggested. On a more optimistic note, BBC 2 will broadcast three documentaries ("Anxiety Attack") on the scare story industry which may herald a more critical approach.

Greenpeace's attack on chlorine continued intensely through the early nineties with a move away from the advocacy of a total ban on chlorine to a focus on individual chemicals such as PVC. In targeting a number of specific chemicals it seems that the green attack lost some of its momentum and the chlorine industry grew in confidence to rebut the attack. Communication with the public about chemical matters, however, remains very difficult.

2 THE CHEMICAL INDUSTRY

The chemical industry and the chlor-alkali industry, in particular, has come a long way from its early days when the safety aspects of the industry were scandalous. We had a long way to go, a hundred years ago the chemical industry was killing its operators. It probably became uneconomic to sustain this death rate, so that the industry began to get its safety act in order. It has certainly achieved this for the last twenty years and we are now at the stage where the chemical plant operator is at a lower fatal risk at work than in his own home. It is only fifty years since our environmental immaturity was demonstrated when, having set up Gossage towers to take hydrogen chloride out of the stinking atmosphere of Widnes and make it into hydrochloric acid, we then poured this into the Mersey and various local waterways. We simply transferred pollution from the air to the water. Though great improvements have been made these are not immediately evident to the public, to them the plume is still coming out of the chimney even if it now carries a fraction of the environmental load.

We who work in the industry are typecast, as are the regulators. We are perceived as having a narrow scientific training, a dependence on logic and a lack of charisma. A majority of us still believe that because such and such is logical to us, then it must be logical to everyone. This overlooks the changed view of the world that scientific training can produce. The physical senses, and common sense, may lead to conclusions that are contradicted by scientific reason. The history of Columbus and Gallileo should show that. Our training in communication accentuates the passive mode of speech. Nobody does or makes anything, it happens or is formed. An awareness of how much is unknown may make us appear unconvinced and unconvincing. Thus we create an image of po-faced boring individuals who communicate in meaningless colourless terms. Fear of criticism has caused the industry members to play their cards close to their chests. This secretiveness has been a breeding ground for suspicion and any popularity from our paternalistic past has plummeted. Given this history it is scarcely surprising that the industry's response to the green attack has been of a low quality and poorly received.

3 PERCEPTIONS OF THE INDUSTRY

3.1 Schools

School children are significantly influenced by green campaigns which use animal protection as a theme and many school teachers have sympathy with Greenpeace views. The overall influence on children is anti-chemicals and in some cases chlorine is singled out for criticism. In general school children do not see the chemical industry as one which will provide a career and are ignorant of its different sections and activities. The chlorine industry has not in the past provided literature to challenge these views.

3.2 Local Communities

Local communities are influenced by the "not in my backyard" syndrome. There is a general distrust of the chemical plant in the community and a suspicion that it is doing something harmful to people. These views will be reflected by town and parish councillors and will generally come strongly to light at times of planning applications. Nobody will get elected today for campaigning to have a chemical works built in the centre of the constituency.

3.3 County Councils

County Councils will take a broader view of the chemical industry and will see it from the socio-economic viewpoint as well as the land use issue. They must consider the use of land already contaminated or blighted by neighbouring activities, employment and infra-structure requirements.

3.4 Partnerships, Agenda 21 Fora

The chemical industry is generally welcomed in partnerships, but more for the possibility of providing funding and resource than as an important member of the community. The resource of technical knowledge can be offered to gain trust among opinion formers and balance the influence of Green Groups. Sponsorship of conservation initiatives both locally and nationally can put company and industry names in the public view in a favouable way while establishing dialogue rather than confrontation with the environmentally concerned.

3.5 Green Groups

Green groups in general view the industry as something society should be without. As in the political arena, calls to abandon 'knocking' the opposition in favour of being constructive receive only lip service as a negative campaign is easier to run than a positive one and will give publicity to the cause. This feature of environmental pressure groups has been noted in the report on the recent oil spillage at Milford Haven.

4 NATURE AT WORK CAMPAIGN

Details of the Chlorine: Nature at Work campaign, have been given in the previous presentation. The value of the campaign in increasing the competence and confidence of the industry in communication is underlined by the preparation of a chlorine advertisement for the London underground. This is a real development from the tentative culture of 1990. This growth in confidence has accompanied Greenpeace's retreat, mentioned earlier, from a complete ban on chlorine, to an acceptance of its use in pharmaceuticals, to an attack on specific compounds such as PVC.

5 RISK/BENEFIT ANALYSIS

Risk/benefit analysis forms an important part of the message to school children arising from the communications committee's work . Here the intention is to encourage a questioning and reasoned approach to matters of safety and the environment. We start with a discussion of the way we think about the plus and minus of chemical manufacture compared with risks taken by each of us in our everyday life. Known benefits of the chlorine industry are raised at this point e.g. water sterilisation, medicinal products and listed against the risks from pesticides and persistent organo-chlorine compounds. To return the arguments to everyday terms the presentation includes the risk/benefit analysis carried out by Mr Bean on a Saturday morning.

> To wash or go dirty? Mr Bean must first decide whether to risk getting soap in his eyes or to spend his day in an unhygienic state. The less evident risks of being dirty are raised.
>
> Use a refrigerator for food or let it rot? Next breakfast, Mr Bean has already made his decision, he has a refrigerator despite its environmental impact because of the convenience and health aspects of fresh food.
>
> Cross the road to the pub or remain thirsty? Being Saturday, Mr Bean decides to quench his thirst at the local pub. This entails crossing a busy road with obvious risks. This is the type of risk we are more willing to take as we believe that we control the situation. As a comedy character the means of control Mr Bean would use might not be as sensible as a pelican crossing.
>
> Have a soft drink, with all those E numbers or a beer? Even before deciding between the multitude of products on display in the pub he must weigh up the risks of alcohol consumption, additives and passive smoking against his thirst, social integration and reduction in heart disease.

This method can be extended (or cut short) as time and the audience allow. We then examine the methods Mr Bean could use to make his decisions, working towards the conclusions that;

1) Decisions are to be based on reason not emotion.
2) That a scientifically valid analysis is essential to the decision-making process.

Finally, the risk/benefit analysis is applied to chlorine in the same way, questioning any assumptions and ensuring the final concensus is based on a reasoned argument.

6 COMMUNICATIONS

Over the last year Jim has asked key people how sound science should be communicated without receiving a satisfactory answer. There have been a few scientists over the years who have communicated effectively with the public. Magnus Pike of hand-waving fame, Patrick Moore with his fixed stare into the camera and his intense rapid-fire talk, and more recently, David Bellamy with his unique style. All three are characterised by delivering their science with passion.

On the question of who the people in Congleton Borough would trust most to communicate with them on safety, health and environmental matters it appears that Greenpeace and David Bellamy are in the lead and the factory inspector and environment agency inspector bringing up the rear in a field of nine candidates. The industry concerned was not placed. People trust those they know, those who have no apparent axe to grind and those that can speak with conviction. To overcome mistrust we must be prepared to speak as people, parents, neighbours, rather than company representatives; to explain what we know and what is unknown in our science and to do so with commitment and passion.

3

HOW A CHLORINE PRODUCER MANAGED TO IMPROVE THE IMAGE OF A MAJOR CHEMICAL PLANT

Mário A. Cilento

Executive Vice-President
Carbocloro Oxypar Indústrias Químicas S.A.
São Paulo - Brazil

1 CARBOCLORO OXYPAR, A STATE-OF-THE-ART PLANT

Carbocloro Oxypar Indústrias Químicas S.A. is a joint venture of Occidental Chemical and Unipar S.A. Its industrial facilities are located in the county of Cubatão, São Paulo, where it produces 235,000 tonnes/year of chlorine (140,000 tonnes/year by the diaphragm process and 95,000 tonnes/year through the mercury process) and 140,000 tonnes/year of EDC.

The plant is world class, employing state-of-the-art technology that merited numerous awards in the region for its performance, particularly in the environmental field.

The Carbocloro plant is also regarded by its international partner Occidental Chemical as one of its top performing plants with regard to conservation and environmental protection. It has been named twice "the best of the group" in the world, while competing with 50 other Occidental sites located in the US and in other countries.

The plant has also received technology awards, such as the De Nora, for the productivity of its anodes that achieved 1,160 tonnes of chlorine per m^2 during eight years of continuous production, while working at current density of 14 ka/m^2.

2 THE SETTING AND THE ENVIRONMENT

In 1955, when Cubatão was chosen as the location for an industrial city and petrochemical center, the reasons seemed obvious: the village of a few thousand people was at the side of the brand new highway that linked South America's largest port, about twenty miles to the east, to the largest market city of Latin America, the city of São Paulo, 50 miles west and 800 meters up the plateau, the capital of the state of the same name.

Actually, it was an inadequate choice. The situation of Cubatão is geographically unfriendly, a place at the bottom of the high slopes of an unstable mountain system, where the pressure of the warm winds from the sea capped by the cold air from the highlands conspire against the dispersion of pollutants.

At that time, there was little public conscience of possible environmental harm. Even the technologies for the protection of the environment were almost non-existent or quite primitive.

With time, the Cubatão industrial center expanded, and with its success, also the village, that emerged from a roadside fishing city into a satellite of the industry, spreading in all directions and getting nearer the plants, to the point of engulfing some of them. Obnoxious side effects such as smells, dust and polluted water were tolerated as a normal price to pay for progress, something to be accepted, much as storms, floods and other natural occurrences.

Then, almost overnight, the world acquired the conscience of conservation and began to cry against the harms caused to the environment by the industrial pollution. Cubatão was then already growing on both sides of the road that links the main highway to the sea resort of Guarujá — the choice beach of the wealthy families from São Paulo. Smells and dust were now an ugly reality to the hundreds of thousands of weekenders that drove down from the plateau every week. In 1976 the government issued the first regulations about environmental protection.

The visibility of Cubatão, because of its proximity to a metropolis of more than 10 million people, caused strong negative public response. In Brazil, as in the rest of the world, NGOs were being created and were exerting pressure on government and communities. The nickname "Valley of Death", that at first was just a fantasy from an inspired newspaper writer, suddenly turned real, when part of a shantytown that had settled over forbidden estates reserved for Petrobras[1] pipelines, burned up, killing a hundred people and raising an uproar in the world press (Figure 1).

No Stars in the "Valley of Death"

A case study in Brazil's unchecked industrial development

Cubatao: 'The Valley of Death'

Jornal soviético noticia Cubatão como o "vale da morte em SP"

ДОЛИНА СМЕРТИ ПОД САН-ПАУЛУ

Ataca la contaminación

Dramática y explosiva degradación del ecosistema terráqueo

Figure 1 *Cubatão was news everywhere. The epithet of "Valley of Death" and the tragedies involving humble people was an assurance of good readership.*

The destruction of the vegetation of the rain forest at the sharp slopes of the Serra do Mar[2] that leads to the São Paulo highlands, was also blamed on the gaseous chemically charged

---

1 The Brazilian government oil monopoly, Companhia Brasileira de Petróleo - PETROBRÁS, the most important Latin American company, has a large refinery in Cubatão from where it supplies almost half of all Brazilian fuel.

2 A mountain system along the litoral of the State of São Paulo, Brazil, that separates the lowlands from the plateau. It is a part of the remaining lush rain forest, that has been severely hit by the woodcutters, palmetto gathers and colonization during the last four hundred years.

effluents of the industry. The industry was later acquitted of most of those claims, but that happened only much later and the industry had to bear the burden of proof.

In 1984, a total of 320 sources of pollution in the soil, water and air were identified and duly classified by the environmental authority, the CETESB (Table 1). This initiative evolved into a formal program for the recuperation of the environment of Cubatão, with the mandatory participation of the 22 industries of the district, as well as the community, the local authorities and the general public.

Table 1 *Sources of pollution identified by the government in Cubatão in 1984*

Sources of Air Pollution	230
Sources of Water Pollution	44
Sources of Soil Pollution	46
TOTAL	320

Although chemical industries are known for their shy presence and perpetual desire for low profile, this time they could not hide. The plan was received with mixed responses, at times with enthusiasm, sometimes reserved, tolerant and often uncooperative. Carbocloro was among those most enthusiastic. At the same time, certain external factors were also conspiring to force a position of the chemical industry face public opinion. The Bhopal mishap, the new language of the Perestroika in the European East and other severe accidents in chemical plants alerted public opinion to the issue of environmental dangers. Meanwhile, in Brazil, the end of the military regime and the new freedom in communications, were circumstances that framed a new scenario. A public relations plan was devised to ensure the needed credibility to the joint effort of the 22 industries, and to provide the public with clear and frank information on how the progressive compliance to the environmental rules were taking place. Slowly, a reversal in the public reaction began to occur.

3 CARBOCLORO, THE EXCEPTION TO THE RULE

For many industries of Cubatão it may have been a painful experience to force a change in their cultures in order to accept and comply with the new model of industrial behavior, that demanded commitments in terms of achievements and deadlines, as well as in transparency when informing the public.

For Carbocloro, however, the new scenario brought no surprises. Its plant in Cubatão was one of the few exceptions to the general rule: a large site, technologically very modern and already demonstrating a conscientious attitude towards environmental problems. The plant of Cubatão, as far back as in 1978, had been obeying the legislation of Brazil and its systems also fitted those of all countries of the Western Hemisphere, not only with regard to environmental protection but also in all matters that had to do with employee safety and industrial hygiene, a policy of never exposing its employees to chemical agents.

Even at that time, when the total investment in the plant was 190 million dollars, already 10%, or 19 million dollars, was applied in environmental control. Carbocloro had already found its own way to position itself as a citizen of the Cubatão community.

Up to 1984, the investments of the Cubatão industries in environmental protection had reached 140 million dollars, but in the following ten years that figure had grown to 700 million dollars, a substantial investment of the 22 companies involved in the project.

From the technical point of view, the program was mostly a success and it could be felt that the majority of the companies really wanted to achieve the goals that had been set by the environmental authorities. By 1993, more than 90% of all sources of pollutants were under control. But the truth is that technology was only one of the factors underlying the program, while the real pusher for results was the enthusiasm generated by the partnership in the form of mutual cooperation, a feeling that was contagious.

4 KEEPING THE TORCH LIT

After some partial objectives were reached with the public relations program, particularly with respect to curbing the media aggressiveness, Carbocloro decided to improve the communications program on its own, free from the constraints of the large group and with much more agility in decisions and in implementing actions. There were special reasons for this attitude.

Carbocloro had been punished by CETESB with an unjust fine originating from an anonymous phone call with accusations that have never been proved true, but still had to pay it and suffer all the publicity consequences. Also, a public opinion survey conducted by the company demonstrated that although Carbocloro always behaved correctly and far beyond the requirements of the law, its image was tainted because it was a part of the quite unpopular chemical business. The fact that its plant had green areas, clean ponds with fishes and was a model of conservation was not being perceived by the general public nor by the community. It was plainly a situation that called for individual and independent action.

5 THE CARBOCLORO INITIATIVE

The strategy devised indicated that Carbocloro should change completely its approach to the community. It would not be a dialogue if the company should only "tell" and the community "listened". A new language was needed to level up with the people of Cubatão. We believed that we would have success if we proved our points by deeds instead of by words, and if we should use as spokespersons people that were already circulating within the community, instead of giving the job to communication professionals that would utilize classical mass reaching techniques.

As a foundation for the Carbocloro project we decided to rely on absolute truth and frankness; and the key for the success was a courageous attitude to face public opinion by opening the plant wide for visitors, in such a way that no other industry ever did before.

6 OPEN PLANT: 24 HOURS A DAY, 365 DAYS A YEAR

Anybody can walk up to the guard at the gate of the Carbocloro Cubatão plant at any hour, and tell him that he wants to visit the plant, even at midnight on a holiday! The guard will find someone to accompany the visitor, somebody who will tell our story with his own words, not parroting a learned script. This was the approach employed by Carbocloro. There would be no restricted areas, nor any restricted information; and if the visitor wanted to take with him a sample of the effluent water from the tubes that feed the Cubatão River he is welcome to do so and the guide will even help him collect it.

7 500 EMPLOYEES: 500 PUBLIC RELATIONS PERSONS

Instead of maintaining professional PR receptionists or guides wearing business suits and affecting rehearsed manners, the plant counts on the good will of 500 voluntary guides, its own employees, all knowing exactly how they feel about the plant and why they were chosen to tell the company message.

Employees are encouraged to show and explain things in their own words, within their own cultural frame, in order to be authentic and convincing, expressing their own personalities, even at the risk of appearing naïve.

8 THE CARBOQUARIUM, A LIVING ARGUMENT

Nothing could be more convincing than the Carboquarium, a large aquarium where living fish can dwell and feed, in a water environment that is the effluent of the plant, just before the liquid is ejected into the Cubatão River.

The idea of the Carboquarium started as a curious experience of an employee, and was immediately adopted by all the workers of the plant. It is now a source of pride, motivation and personal responsibility to everyone.

The Carboquarium concept encompasses both medium and message, in a format that otherwise would demand hundreds of technical pages to explain. Simple, common people, particularly schoolchildren, are convinced on the spot when they see the nimble creatures behaving normally. It is a live document of the efficiency of the control systems.

Figure 2 *The Carboquarium is maintained by the employees themselves, that consider it a symbol of their own performance in conservation of Nature. Many varieties of fish can live in an environment of the treated effluent from the plant.*

9 STRENGTHENING THE LINKS WITH THE COMMUNITY

Other communications and social activities complement the contact of the plant with the community. Throughout the year Carbocloro sponsors and hosts educational activities, such as ecological contests, that offer prizes to the best performances.

A competition among the local schools involved 16,000 students annually and the contests are held in the plant premises. While chosen groups of students are participating in the matches, professors and other groups of students visit the industrial site and tour the outdoors where there are botanical gardens and ponds (Figure 3) with specimens of the rich fauna and flora of the region, waterfalls and the orchard with fruits of the season.

The Carbocloro plant outdoor space is sufficiently large and natural to accommodate other activities. It has already served as grounds for boy scouts' camping for four days, an initiative that proved to be an excellent manner to secure the participation of the community children in contact with the natural environment of the slopes of the Serra do Mar rain forest, under ideal conditions of safety. These children are the best ambassadors that Carbocloro would ever desire to have, acting on its behalf within the community.

When the industrial community adopted the APELL, a program based on the recommendations of UNEP, for alerts and preparation of the neighborhood for eventual tecnological emergencies, Carbocloro was one of the active leaders in the training of the Cubatão community. The program today is incorporated into the activities of the Brazilian Chemical Industries Association - ABIQUIM and is being implemented all over Brazil.

Figure 3 *Visitors, particularly the young ones, delight in watching the healthy specimens of birds and small mammals at the Carbocloro gardens. Seeding and replanting of saplings of vegetal species that are natives of the Serra do Mar rain forest are also grown right at the Carbocloro's greenhouse.*

10 BEFORE AND AFTER - HOW OPINIONS CHANGE

The findings of a public opinion survey in Cubatão made in 1986 demonstrated that the program of visits to the plant had immediate and fully convincing results. From among people that had negative opinions about Carbocloro the rate of opinion change was overwhelming.

Over 30,000 people from all parts of Brazil have visited the Carbocloro plant from 1985 to-date and each visitor is requested to express his opinion on the image of the company, before and after the visit. The questions deal not only with environmental points but also with general impressions, even the attention and courtesy of the employees. From the inception of the program (Table 2), the trend demonstrated the value of the visits.

The same survey was applied throughout eleven years of the Open Plant Program and the results were surprising. The prevailing impression about Carbocloro reached such a point that the opinion about the company is now already quite positive even before people visit the plant (Table 3).

From the tables 2 and 3 below, we concluded that the sum of "bad" and "regular" impressions of people before they visited the plant improved from 49% down to 32%, showing that word-of-mouth communication is contributing to the continuously improving image of the company before the general public.

But real improvement can also be noticed in the "good" and "excellent". Not only the opinions migrated from "bad" and "regular" to a positive position, but also migrated from "good" to "excellent". The position in 1986 (Tables 4 and 5) shows the change.

Table 2 *Opinions that considered Carbocloro as BAD or REGULAR – 1986*

	BAD	*REGULAR*
Before the visit to the plant	9%	40%
After the visit to the plant	0	1%

Table 3 *Opinions that considered Carbocloro as BAD or REGULAR - 1997*

	BAD	REGULAR
Before the visit to the plant	3%	29%
After the visit to the plant	0	0%

Table 4 *Opinions that considered Carbocloro as GOOD or EXCELLENT - 1986*

	GOOD	EXCELLENT
Before the visit to the plant	36%	12%
After the visit to the plant	19%	79.5%

In 1997, even the "good" migrated to "excellent", as can be seen in Table 5.

Table 5 *Opinions that considered Carbocloro as GOOD or EXCELLENT - 1997*

	GOOD	EXCELLENT
Before the visit to the plant	50%	14%
After the visit to the plant	8.9%	91%

At this time, the program can surely be considered as a winner, and all the measurements show a very high level of acceptance of the company in the community.

11 ACCEPTANCE OF CARBOCLORO IS THE ACCEPTANCE OF THE CHLORINE INDUSTRY

But, probably more important than the results obtained from the program by Carbocloro, a much more encompassing positive consequence is being felt. The community does not fear the chemical industry anymore, nor objects to the presence of chlorine plants in their neighborhood.

We are sure that we destroyed prejudices against chlorine and we brought transparency and assurance to our community, where fear and distrust existed before.

We feel that we are doing something special for the image of the whole chemical industry and particularly for the chlor-alkali sector.

For this reason, Carbocloro is strongly convinced that its pioneer program that worked for Cubatão is certainly feasible to be adapted to other companies, even in places with different cultural formation, in other communities around the world. The principles that guided the Carbocloro program — based on total openness and frankness in dealing with the public and respecting their intelligence and their right to be informed as citizens — are universal.

As expressed by Congressman and former mayor Oswaldo Justo, when he registered his personal impressions on the visitor's book at the Carbocloro plant: *"...As I visit Carbocloro, I believe that it is possible for the charms of Nature to coexist along with high technology, even when involving products that are potentially dangerous. I see that as with business results, the company is concerned with flowers, birds, plants and all things that are beautiful and created by God. Congratulations! Everything here is as transparent as the "Carboquarium...".*

12 "THE HOUSE OF CHLORINE"

The Carbocloro plant today is the "House of Chlorine", a real house with its door permanently open to the community and for everybody to come in and look. It is the proof that friendly and frank chatting, in mouth-to-mouth communications with neighbors can work better than elaborated administrative systems.

The chlorine plant in Cubatão is an industrial citizen that can be trusted by the people of the neighboring cities, and is a respected business that makes products needed by all people; and does it in a manner that is reliable and friendly.

4

BEING GOOD AND BEING SEEN TO BE GOOD: A MATTER OF PERCEPTION

Samantha Benoy

ICI Halochemicals
PO Box 13, The Heath
Runcorn, Cheshire WA7 4QG, UK

1 INTRODUCTION

'Perception is reality' is one of the conclusions of research conducted by leading American risk communication expert, Professor Vincent Covello.

The way in which we understand and interpret our world depends on a variety of factors; upbringing, education, social and economic conditions, peers and is, of course, heavily influenced by hundreds of competing communicators sending out messages through a myriad of media channels. Every day we are bombarded with complex information which we either interpret for ourselves or seek clarification from others through radio, television and newspaper commentary. There is no doubt that our perceptions are formed and heavily influenced by communication.

In particular, we are influenced by sources which we believe to be credible. For example, young children will take on trust the instruction of a favourite teacher; as we mature, we seek-out interpreters of our world who largely reflect our own personal views and re-inforce them - and who, to us, are credible. This holds true as much for the choice of newspaper we read as for which colleagues we go to for advice.

Those of us working in the chlorine industry know how much it contributes to our quality of life and to society. We know how seriously we take our responsibilities towards the community and the priority we give to safety, health and the environment. But sometimes we forget that the public does not necessarily know about or understand who we are and what we do. And do we really know our stakeholders?

2 AN INSIGHT INTO PUBLIC PERCEPTION

To find out more about public perception, in March 1996, the Chemical Industries Association's chlorine communications committee commissioned a survey by Market & Opinion Research International (MORI). A nationally representative sample of 1,028 adults aged 15 and over were interviewed face-to-face at 74 sampling points throughout Britain. The questionnaire repeated the structure of an earlier piece of research conducted in November 1991.

The first point made in the summary is "Public attitudes to chlorine continue to be broadly negative. The proportion considering it to be harmful to the environment (41%) remains over twice that considering it to be safe". More alarming still is the finding that two in five believe that the environmental harm of chlorine is greater than the advantages it bestows. A further interesting point is that "the familiarity of the public with the chlorine industry remains very low" - around 96% of the sample felt they knew little or nothing about us - "and those that are familiar tend to be more negative towards it" which suggests that the information they do receive is not generated by industry.

So, this is one insight into public perception of the chlorine industry. Let's unravel where these views may have originated. Environmental groups spend a significant amount of money on communication; they are masters of the media. Not just the traditional channels of television and newspapers but of the internet and sophisticated communications technology. Quick to respond to opportunities and threats, environmental groups are supported by volunteers who ensure that the current campaign is cascaded from national issue to a topic of immediate local concern e.g. dioxins. From "Devil's Element" to the more recent publication "Taking Back our Stolen Future", environmentalists have been active against chlorine. It is interesting to note that while membership of Greenpeace is said to be falling, our research indicated a stronger overall support for the activities of pressure groups.

3 OPPOSING INFLUENCE

In addition, pressure groups are trying to make themselves more acceptable to a wider audience with "solutions campaigning" or in our terms product deselection. A good example of this took place on the chlorine derivatives stage. A company decided to have a high profile launch of an aqueous cleaning system directly challenging the use of perchloroethylene in drycleaning operations. Not only did popular celebrity scientist, Dr David Bellamy, give it his backing but Greenpeace assisted the company to market the product by endorsing it and using hallmark tactics such as models wearing gas masks to indicate the problems caused by using per for drycleaning. The story made the national press - though only in snippets - and also some local papers. We were able to provide the other side of the story for the consumer magazine 'Good Housekeeping' (readership in the UK and Eire of 2,382,000) and ICI ran a successful seminar "Perchloroethylene in Perspective" and fielded a spokesman in a shared-platform debate with Greenpeace at a UK Textile Services Association meeting in the autumn.

A further example, and potentially rather more important in its ramifications for the chlorine industry is Greenpeace's targeting of retailers and specifiers with a high profile campaign against PVC. Starting last year, an elaborate communications effort to promote alternatives to PVC in all applications is likely to continue over the next few years and it is essential for the chlorine and PVC industries to work together presenting a united front and consistent communication. While individual companies already have strong relationships with their customers, the need to collaborate more strongly at trade association level became evident in 1996 and is a key aspect to this year's activity.

4 INDUSTRY'S POSITION

The gap between the public's perception of the role of chlorine in society and the reality of the benefits it bestows on society is very obvious. Our "Responsible Care" message is not getting across and we need to find more immediate ways of communicating without an overload of scientific facts which many people find difficult to assimilate. In today's world, it is not enough for industry to be good in terms of safety, health and environmental performance, contribution to national and local economies, and to contribute to the high standards of living to which we have become accustomed. We also need to be seen and heard to be good.

For the past five years, the Chemical Industries Association's chlorine sector group has invested in a communications committee which comprises representatives from The Associated Octel Company, Dow Chemical Company, Elf Atochem UK, Hays Chemicals, ICI Chlor-Chemicals, Milennium Inorganic Chemicals (formerly SCM Chemicals) and Rhone-Poulenc Chemicals. We aim to provide credible, easily understood, communication about chlorine. To influence perceptions we want to ensure people understand our industry and its vital role in society, perceive it to be of tremendous value to everyday life and have a favourable overall opinion of chlorine.

While we aim at specific target audiences, one of our major challenges is in reaching the end consumer who does not recognise the value of chlorine - except in water treatment and household bleach. Environmental pressure groups, most notably Greenpeace, have channelled considerable resources into striking us where we are most vulnerable - with consumers. We know from CIA chemical industry research that the chemical industry as a whole has a low credibility factor and that environmental pressure groups fare better than commercial management in the credibility stakes.

5 THE STORY SO FAR...

Through the CIA's chlorine sector group, a wide range of excellent presentation materials has been produced over the past five years. Outlining the issues in a very accessible way, a major dossier of overheads has been compiled to cover all the needs of the chlorine industry supporter active in the community. To support this, Chlement, the Ecuadorian tree-frog which produces a natural organochlorine compound lends a webbed foot to promotional activities putting his image to pens, fridge magnets, mouse (or is that frog) mats, PVC carrier bag, t-shirts, balloons and a leaflet. If it's the more detailed response to current issues you are looking for then we have a dossier covering major topics.

In common with all other industries, we do not have an automatic right to the trust and support of our stakeholders. Rather we must create respect through two-way communication, listening as well as speaking and being seen to do what is right. Talking to people provides an opportunity to understand the needs and concerns of those who can be affected by our activities and how they perceive environmental issues. Apart from being able to bring information back into our organisations from outside, it is also a unique opportunity to talk about the benefits of an industry we support - without being defensive. We've got good material, let's make sure we use it.

6 NEW DEPARTURES

Working closely with Euro Chlor our continued emphasis is on communicating confidently and credibly to key audiences, creating opportunities to express our views and ensure that we express our messages in such way that people feel they can buy into them. This will help us to achieve industry credibility, build relationships and stimulate interest in the benefits of chlorine. Our choice of primary target audiences this year reflects our concern about the concerted attack on PVC and also the need to reach beyond our customers to our customers' customers and even further down the supply chain to the ultimate consumer.

One relatively new organisation which is set to have a dramatic impact on people's perceptions of Greenpeace activities is the Chlorophiles. A voluntary group of employees from the chlorine and PVC industry, the Chlorophiles put across a strong message effectively; for example by demonstrating against Greenpeace in Dusseldorf, and through publications such as "The Hidden Face of Greenpeace". The Chlorophiles also turned up with Greenochio - a large puppet with a long nose covered in ivy (to represent the lies of Greenpeace) outside the UK venue for Greenpeace's conference "Brent Spar and After". Anyone can find out more about the Chlorophiles by accessing their internet site.

7 TOWARDS A BALANCED VIEW

We have all been in a situation where we have needed to find a reputable supplier of a product or service and, of course, we ask our friends or colleagues for a recommendation. Our perception of the suitability of a supplier is often based on the advice of experts, for example internal purchasing managers. While we consult the sales literature, there is a distinct feeling of "they would say that wouldn't they". Not surprisingly, we prefer an unbiased opinion. This is a key consideration for us in the chlorine industry, if we are to build credibility we need high profile third party endorsement - for example, from our customers, from journalists and academics. In short, opinion formers.

Increasingly, we are learning the importance of a balanced view. When we commissioned a schools' resource pack for 14-16 year olds, the product was one which did not try to "sell" the industry but rather tried to educate about risk and benefit setting chlorine in its complex societal context. This year, there will be a book aimed at younger children telling the story of chlorine, again from an objective view point.

And what of those people who take facts and interpret them for others - the opinion formers? Increased contact will help us to build relationships and ensure accurate information forms the basis of their interpretation. Last year we sent information to more than 50 freelance consumer journalists and spoke to 39 staff writers at consumer magazines to assess interest in a chlorine story. A reflection of how chlorine has slipped down the issues agenda is that none of them appeared aware of the chlorine debate - a good opportunity, therefore, for us to give a positive message about the value of our product.

A good example of an organisation which will lend balance to the emotional versus scientific debate is the European Science and Environment Forum (ESEF). The Forum boasts a wide range of eminent writers including Dr John Emsley who wrote a thought-provoking piece in 'The Independent' last year highlighting how BSE may have been avoided if methylene chloride had not been classified a possible human carcinogen. As we move forward into the era of risk communication, scientific fact will need to be conveyed in a straightforward and credible manner.

8 CLOSER ALLIANCES

As the year progresses, we focus increasingly on our strategic alliances. At a trade association level we have excellent contact with the PVC industry and the challenge is for all chlorine producers and users to ensure that our customers have ready access to information and materials which put the record straight for chlorine.

The chlorine and PVC industries employ thousands of people in the United Kingdom. If every one of us has a clear understanding of the benefits of our product and shares that knowledge with just six people in another industry or in the local community, just imagine what the cumulative effect would be. We need your help and we can provide you with the materials you need to become a supporter of the chlorine industry in the battle against "don't know, don't care" - we do know and we should care enough to act.

5

BROMATE AND CHLORATE - EVALUATION OF POTENTIAL EFFECTS IN AQUATIC ORGANISMS AND DERIVATION OF ENVIRONMENTAL QUALITY STANDARDS.

Thomas H. Hutchinson[1] and Dolf J. van Wijk[2].

[1]Brixham Environmental Laboratory,
ZENECA Limited,
Brixham, Devon TQ5 8BA, UK and
[2]Akzo Nobel Central Research,
PO Box 9300,
6800 SB Arnhem, The Netherlands.

1 INTRODUCTION

Bromate (BrO_3^-) and chlorate (ClO_3^-) have not been reported to occur naturally in surface waters. It is known, however, that both compounds may enter the aquatic environment as a result of various anthropogenic activities. These activities include the chlorination of power plant cooling waters, the disinfection of drinking waters using chlorine, ozone or perozone and the use of sodium hypochlorite as an industrial and domestic disinfectant.[1,2] Given the potential for bromate and chlorate to enter aquatic ecosystems, a critical review has been undertaken of the available data on the effects of these compounds on aquatic organisms. Based on computer literature searches, and a variety of review articles, original reports were obtained wherever possible for critical review. Where there were gaps in the available information or inconsistencies regarding the conclusions reached in various historical studies, additional experimental studies were conducted with a view towards resolving these uncertainties; full details of these studies are described elsewhere.[3,4] This overview summarises the critical aquatic toxicity values for bromate and chlorate, with the aim of assisting in the subsequent derivation of relevant environmental quality standards (EQS values), according to local and regional circumstances.

2 MATERIALS AND METHODS

2.1 Literature Review

Data on the effects of bromate and chlorate on aquatic organisms were gathered from a wide range of sources, including a variety of electronic databases, peer-reviewed publications and technical reports. The original publications were reviewed wherever possible to properly interpret the original data and associated experimental methods. This was especially important, for example, where there was an apparent conflict in the published data on the effects of bromate on oyster embryo development.[4]

2.2 Analysis of Aquatic Toxicity Data

The published data, together with the new experimental information, were collated and reviewed according to the following procedure. Where possible, data were cited using the following standard aquatic toxicity endpoints, including the median effect (lethal) concentration or E(L)C50 for a given exposure period, the Lowest Observed Effect Concentration (LOEC) and the No-Observed Effect Concentration (NOEC). In some of the older publications, effects are expressed as the median tolerance limit (TLm or TL50), however, such values were interpreted as being equivalent to the LC50 for the purpose of this review. Additionally, some workers use the term 'toxicity threshold' for bromate, this endpoint being considered to be approximately equivalent to the LOEC.[5]

3 RESULTS

Published data were collated according to the major taxonomic groups of aquatic organisms. All values cited represent the concentration of bromate or chlorate ion. Studies were classified as acute when the exposure period was only a few days duration and was limited to one lifestage of the organism. In turn, chronic studies included multi-generation exposure periods or where the exposure period included a substantial part (ca. $^{1}/_{3}$) of the time to sexual maturity of the organism in question.[6]

3.1 Bromate Data

There appear to be no data available on the effects of bromate on freshwater algae or higher plant species, although data are available relating to effects in marine phytoplankton (Table 1). In summary, Erickson and Freeman[7] studied the effects of sodium bromate ($NaBrO_3$) on four species of marine phytoplankton, using cell division (growth) of the cultures as the test endpoint. For the species *Glenodinium halli* and *Thalassiosira pseudonana,* compared with the control cultures, cell growth was increased in the presence of 13.6 mg BrO_3^- l^{-1} (unspecified exposure time). For all species examined, the EC25 and EC50 values (unspecified exposure time) were both >13.6 mg BrO_3^- l^{-1}.

There is a limited range of data available on the effects of BrO_3^- on freshwater and saltwater invertebrate species. In terms of fresh water species, Anderson[8] reported a 48 h LC50 of 179 mg BrO_3^- l^{-1} for the water flea, *Daphnia magna,* while Jones[9] reported a 48 h LC50 of 2258 mg BrO_3^- l^{-1} for the planarian ('flatworm'), *Polycelis nigra.* The available saltwater invertebrate data are limited to a range of crustacean and bivalve mollusc species. For the most sensitive marine crustaceans reported, Crecelius[10] cited a 24 h LC50 of 176 mg BrO_3^- l^{-1} for adult mysid shrimps. For the bivalve molluscs, the data from various authors are somewhat contradictory and effects on oyster embryo-larvae have been reported over at least three orders of magnitude.[10,11] Recent oyster embryo development studies have attempted to resolve these differences, indicating a 24 hr EC50 value of 170 mg BrO_3^- l^{-1}.[4] These key values are summarised in Table 1.

For fish, all of the reported data on the effects of BrO_3^- on fish relate to estuarine or marine species. For juvenile fish (the life stage most commonly used in aquatic toxicity tests), the 96 h LC50 values were 512 mg BrO_3^- l^{-1} for chum salmon (*Oncorhynchus keta*) and 427 mg BrO_3^- l^{-1} for spot (*Leiostomus xanthurus*). In addition, the 10 d LC50 for juvenile spot was observed to be 279 mg BrO_3^- l^{-1}.[10, 12] The embryo-larval stages of fish are traditionally regarded as being the part of the life cycle most sensitive to toxicants and, therefore, data on the effects of bromate on these life stages are of particular importance. Richardson *et al*[12] studied the effects of bromate on striped bass (*Morone saxatilis*) and observed a 24 h LC50 of 697 mg BrO_3^- l^{-1} based on embryo hatching rates. In contrast, the newly hatched fish larvae were markedly more sensitive to bromate, with a 96 h LC50 value of 31 mg BrO_3^- l^{-1}. However, it was observed in the same study that somewhat older larvae (starting from 4 days post-hatch, termed 'pro-larvae') were significantly less sensitive to bromate. A summary of these data is given in Table 1.

3.2 Chlorate Data

In comparison with bromate, there is a greater amount of published data available on the effects of chlorate on aquatic organisms. This may reflect historical concern over the potential environmental side effects of sodium chlorate when this compound is used for weed control. This paper summarises the key data available on the aquatic toxicity of chlorate and is based upon a more extensive data review recently published by van Wijk and Hutchinson.[3]

The large number of data available have been condensed into a summary of the critical values (Table 2). This summary takes into account a number of the important experimental criteria (e.g. quality of test method, results of more recent repeated studies, etc.) as has been discussed in detail elsewhere.[3] Based on these data, it appears that marine macro brown algae are exceptionally sensitive to chlorate, with a six month study in *Fucus vesiculosus* from the Baltic Sea indicating an LOEC of 0.015 mg ClO_3^- l^{-1}.[13] In contrast, under comparable environmental conditions (e.g. low salinity), marine macro red algae were insensitive to chlorate.

Based on the available data, chlorate therefore appears to be non-toxic (acute toxicity >100 mg ClO_3^- l^{-1}) to most of the freshwater and marine species examined. However, chlorate is highly toxic (acute toxicity < 0.1 mg ClO_3^- l^{-1}) to certain macro brown algae species tested under low salinity and nitrate limited conditions (the NOEC after 6 months was reported to be approximately 0.005 mg ClO_3^- l^{-1}).

4 DISCUSSION

The potential impact of bromate and chlorate on aquatic ecosystems has been considered in view of on-going technical developments in the chlor-alkali industry and with respect to new legislative developments. As both bromate and chlorate may enter the aquatic environment from several industrial and domestic sources, it is necessary to evaluate the hazardous properties of these substances in aquatic ecosystems for the derivation of local and regional water quality standards.

For bromate, studies with a limited number of freshwater species indicate that the critical aquatic toxicity value currently available is 179 mg BrO_3^- l^{-1} (48 h LC50 to the water flea, *Daphnia magna*) (Table 1). For saltwater (estuarine and marine) species, the lowest toxicity value is 31 mg BrO_3^- l^{-1} (96 h LC50 to newly hatched striped bass larvae) (Table 1). It should be recognised, however, that there is a paucity of data on the potential effects of bromate on freshwater species, including algae and fish. Consideration should therefore be given to evaluating the toxicity of bromate to these groups of organisms in order

Table 1 *Bromate - summary of critical values for aquatic toxicity*

Test organism	Lifestage	Test endpoint	Exposure period	Effect concentration (mg BrO_3^- l^{-1})	Reference
Microorganisms: no data available					
Freshwater plants: no data available.					
Estuarine and marine plants:					
Microalgae	Lifecycle	EC25 (growth)	?	>13.6	7
Freshwater invertebrates:					
Daphnids	Juvenile	LC50	48 h	179	8
Estuarine and marine invertebrates:					
Mysid shrimp	Adult	LC50	24 h	176	10
Pacific oyster	Embryo	EC50	24 h	170	4
Freshwater fish: no data available.					
Estuarine and marine fish:					
Spot	Juvenile	LC50	4 d	427	12
Spot	Juvenile	LC50	10 d	279	12
Striped bass	Larvae	LC50	96 h	31	12

to strengthen the basis for environmental hazard assessment and in developing environmental quality standards.

For chlorate, there is a far more extensive database from which to identify critical aquatic toxicity values. Based on the data summarised in Table 2, it is clear that there are marked species differences in chlorate toxicity, with marine macro brown algae being especially vulnerable. The sensitivity of these marine macro brown algae to chlorate suggests that the compound may be acting via specific biochemical processes unique to this group of organisms. Since the only reported mechanism by which chlorate is phytotoxic is via the nitrate reductase pathway, it is possible that the high toxicity of chlorate to the marine macro brown algae is related to this aspect of plant biochemistry. In brief, nitrate reductase is normally active in plants utilising nitrate as the primary source of nitrogen. It is hypothesised that there may be important differences in the level of activity of this enzyme according to endogenous (e.g. inter-species differences) and exogenous (e.g. ambient nutrient levels) factors. In particular, there appears to be a close association between chlorate toxicity in plants and the available sources of nitrogen (namely, N-NO_3 or N-NH_4) (reviewed in reference 3).

Bearing these considerations in mind, studies of chlorate toxicity in a number of

Table 2 *Chlorate - summary of critical values for aquatic toxicity*

Test organism	Lifestage	Test endpoint	Exposure period	Effect concentration (mg ClO_3^- l^{-1})	Reference
Microorganisms					
Microtox™	-	EC50 (bio-luminescence)	15 min	34,510	14
Freshwater plants:					
Chlorella vulgaris	Lifecycle	LOEC (growth)	48 h	334	15
Estuarine and marine plants:					
Microalgae (*Phaeodactylum tricornutum*)	Lifecycle	NOEC (biomass)	72 h	50	16
Macroalgae (*Fucus vesiculosus*)	Adult	NOEC (frond growth)	6 mon	0.005	13
Freshwater invertebrates:					
Daphnids	Juvenile	NOEC (survival)	24 h	600	17
Estuarine and marine invertebrates: no data available					
Freshwater fish:					
Cherry salmon	Larvae	NOEC (survival)	72 h	35	18
Estuarine and marine fish: no data available					

freshwater species indicate that the critical aquatic toxicity value currently available is 35 mg ClO_3^- l^{-1} (72 h NOEC for cherry salmon, cited in Table 2).[18] For saltwater (estuarine and marine) species, the lowest toxicity value is 0.005 mg ClO_3^- l^{-1} (6 month NOEC for macro brown algae, cited in Table 2).[13] Under other circumstances, however, the available data suggest that the critical toxicity value to microalgae is approximately 50 mg ClO_3^- l^{-1}. According to the environment of concern, these values may be used to calculate acceptable levels of chlorate in receiving waters. Importantly, it is also concluded that an improved understanding of the actual mode of toxicity in sensitive plant species is desirable in order to help improve the environmental hazard assessment of both bromate and chlorate. In order to strengthen the scientific basis for establishing acceptable levels of these compounds under different environmental conditions, such studies should also examine the potential modulatory role of ambient nutrient levels (viz. ammonium and nitrate) and salinity on the expression of bromate or chlorate toxicity in aquatic plants.

5 ACKNOWLEDGEMENTS

The authors thank Mr. S. Ingleby (ICI Chemicals and Polymers, Runcorn, Cheshire, UK) and Mr. J. Wierink (Akzo Nobel Chemicals, Business Unit Base Chemicals, Amersfoort, the Netherlands) for their support towards this review. Our thanks also to colleagues in Akzo Nobel, Arnhem and ZENECA, Brixham for their valuable technical assistance.

6 REFERENCES

1. D.T. Burton and L.B. Richardson. An investigation of chemistry and toxicity of ozone produced oxidants and bromate to selected estuarine species, US EPA, EPA-600/4-81-040, 1981.
2. L.B. Richardson, D.T. Burton, G.R. Helz and J.C. Rhoderick. Residual oxidant decay and bromate formation in chlorinated and ozonated sea water. *Water Research,* 1981, **15**, 1067-1074.
3. D.J. van Wijk and T.H. Hutchinson. The ecotoxicity of chlorate to aquatic organisms: a critical literature review. *Ecotoxicology and Environmental Safety,* 1995, **32**, 244-253.
4. T.H. Hutchinson, M.J. Hutchings and K.W. Moore. A review of the effects of bromate on aquatic organisms and toxicity of bromate to oyster (*Crassostrea gigas*) embryos. *Ecotoxicology and Environmental Safety,* in press.
5. G.M. Rand and S. Petrocelli. Fundamentals of Aquatic Toxicology. Hemisphere Publishers, 1985, pp 666.
6. ECETOC. *Aquatic Toxicity Data Evaluation.* Technical Report number 56. European Centre for Ecotoxicology and Toxicology of Chemicals, Brussels, Belgium, 1993, pp 66.
7. S.J. Erickson and A.E. Freeman. Toxicity screening of fifteen chlorinated and brominated compounds using four species of marine phytoplankton. In: Water Chlorination: Environmental Impact and Health Effects, Volume 2 (eds., R.L. Jolley, H. Gorchev and D.H. Hamilton), 1978, pp 307-310. Ann Arbor Science.
8. B.G. Anderson. The toxicity thresholds of various sodium salts determined by the use of *Daphnia magna. Sewage Works Journal*, 1946, **18**, 82-87.
9. J.R.E. Jones. A study of the relative toxicity of anions, with *Polycelis nigra*, as a test animal. *Journal of Experimental Biology*, 1941, **18**, 170-181.
10. E.A. Crecelius. Measurements of oxidants in ozonated sea water and some biological reactions. *Journal of the Fisheries Research Board of Canada*, 1979, **36**, 1006-1008.
11. M.E. Stewart, W.J. Blogoslawski, R.Y. Hsu and G.R. Helz. By-products of oxidative biocides: toxicity to oyster larvae. *Marine Pollution Bulletin*, 1979, **10**, 166-169.

12. L.B. Richardson, D.T. Burton and J.C. Rhoderick. Toxicity of bromate to striped bass ichthyoplankton (*Morone saxatilis*) and juvenile spot (*Leiostomus xanthurus*). *Journal of Toxicology and Environmental Health*, 1981, **8**, 687-695.
13. A. Rosemarin, K. Lehtinen, M. Notini and J. Mattson. Effects of pulp mill chlorate on Baltic sea algae. *Environmental Pollution*, 1994, **85**, 3-13.
14. Macauley, A.O. (unpublished data).
15. L.P. Solomonsson and B. Vennesland. Nitrate reductase and chlorate toxicity in *Chlorella vulgaris* Beijerninck. *Plant Physiology*, 1972, **50**, 421-424.
16. T.H. Hutchinson. Sodium chlorate: toxicity to the marine diatom *Phaeodactylum tricornutum*. ZENECA Brixham Environmental Laboratory report BLS 1741/B, 1994.
17. G. Bringmann and R. Kühn. Befunde der Schadwirkung wassergefahrdender Stoff gegen *Daphnia magna*. *Wasser Abwasser-Forsch.*, 1977, **10**, 161-166.
18. Y. Matida, Y. Furuta, H. Kumada, H. Tanake, M. Yokote and S. Kumura. Effects of some herbicides applied in the forest to freshwater fishes and other aquatic organisms: II. Effects of sodium chlorate and ammonium sulfamate to the aquatic organisms in an artificial stream. *Bulletin of the Freshwater Fisheries Research Laboratory*, 1975, **25**, 55-62.

6
ELECTROCHEMICAL AND SPECTROSCOPIC STUDIES OF RUTHENIUM-BASED CHLORINE ANODES

L.M. Peter, A. Shingleton and R.C. Walker
School of Chemistry, University of Bath, Bath BA2 7AY

D. Hodgson
ICI Chemicals and Polymers, Research and Technology Department
ETB Technological Centre, PO Box 9, Castner Kellner, Runcorn, Cheshire WA7 4JE

1 INTRODUCTION AND OVERVIEW OF TECHNIQUES

The work described in this paper forms part of an ongoing collaborative research programme involving the University of Bath and ICI Chemicals and Polymers (ETB Technological Centre)[1]. A primary objective of the programme has been to develop a range of experimental techniques that could be applied to the evaluation of fundamental and applied aspects of anode coating and performance.

The electrochemical behaviour of mixed Ru/Ti oxide electrodes has been investigated by a range of electrochemical methods as well as by *in-situ* reflectance spectroscopy. The techniques and the type of information that they provide are summarised in Table 1. In this paper we concentrate attention on the last three methods.

Electrochemical impedance spectroscopy (EIS), sometimes referred to as *ac* impedance, involves application of a sinusoidal voltage or current to the electrode. In the present study of chlorine anodes, the perturbation involved superimposition of a sine wave current on the *dc* current in the chlorine evolution region. Analysis of the frequency dependence of the resulting *ac* component of the electrode potential gives information about the Tafel slope and the electrode capacitance under conditions relevant to industrial chlorine generation[2,3].

Table 1. ***Experimental Techniques***

TECHNIQUE	INFORMATION
cyclic voltammetry	surface roughness, double layer capacitance
iR corrected steady-state i-V curves	non-ohmic overpotentials
electrochemical impedance spectroscopy	catalytic activity, Tafel slope, double layer capacitance
open circuit potential decay after current interruption	mechanism of chlorine evolution, catalytic activity, surface adsorption
potential modulated uv/visible reflectance spectroscopy	information about oxidation states of ruthenium

The analysis of open circuit potential decay (OCPD) measurements[4-6] following current interruption offers insights into the kinetics and mechanism of chlorine evolution. The technique involves interruption of the current and subsequent recording of the potential decay with a fast ADC/computer system. The main advantage of the method is that it provides information under conditions corresponding to the current densities used with commercial chlorine anodes, but eliminates the ohmic overvoltage component which dominates the steady state current-voltage characteristics.

Our most recent work uses potential modulated uv/visible reflectance spectroscopy (PMR)[7] to detect the changes in absorption/reflection that accompany oxidation and reduction of ruthenium species. The technique is based on application of an *ac* potential perturbation to the electrode and detection of the corresponding small modulation of the intensity of a monochromatic light beam reflected from the electrode surface. PMR spectra can be related to the presence of different Ru species, and the potential dependence of particular species can be followed by fixing the wavelength and scanning the potential.

2 EXPERIMENTAL

Two types of working electrode were used. The first was fabricated from roughened titanium strip coated by a mixed Ru/Ti oxide layer produced by a conventional spray-bake method. These electrodes were used for the EIS and OCPD experiments. The area to be examined was restricted to 1 cm^2 on one side by coating the electrode with Apiezon wax, the only sealing material found to be adequately robust for measurements in chlorine-saturated brine at elevated temperatures. A large area RuO_2/TiO_2 coated strip was used as secondary electrode in the chlorine evolution measurements, and a reversible chlorine electrode was used as a reference electrode[8]. The second working electrode used for the PMR experiments was a smooth ruthenium-plated platinum disc. Highly reflective adherent Ru films with thicknesses in the micron range were obtained by plating from an aqueous solution of the ruthenium(IV) nitro bridged complex $K_3[Ru_2NCl_8(H_2O)_2]$ (Johnson Matthey) at 70° C. A platinum secondary electrode and a saturated calomel reference electrode were used.

Electrochemical impedance spectroscopy was performed under constant *dc* current conditions using a Solartron 1280 electrochemical interface and 1250 frequency response analyser in conjunction with ZPLOT and ZVIEW software packages. Open circuit potential decay curves and ohmic drop corrected i-V plots were recorded using a mercury-wetted relay to interrupt the cell current. Potential decays were recorded with reference to a reversible chlorine electrode using a RISC based computer with a fast analogue to digital converter, and the decay curves were fitted to the theoretical overpotential time expressions for different mechanisms.

The PMR set-up is shown in Figure 1. The electrochemical cell is equipped with quartz windows through which a beam of monochromatic light is reflected from the electrode surface and focused on a silicon photodiode with enhanced uv response. The output from the photodiode amplifier is connected to a digital lock-in amplifier which also provides the sinusoidal modulation signal to the potentiostat. The system (monochromator and lock-in amplifier) is controlled by a PC and is capable of resolving $\Delta R/R$ down to 10^{-5}.

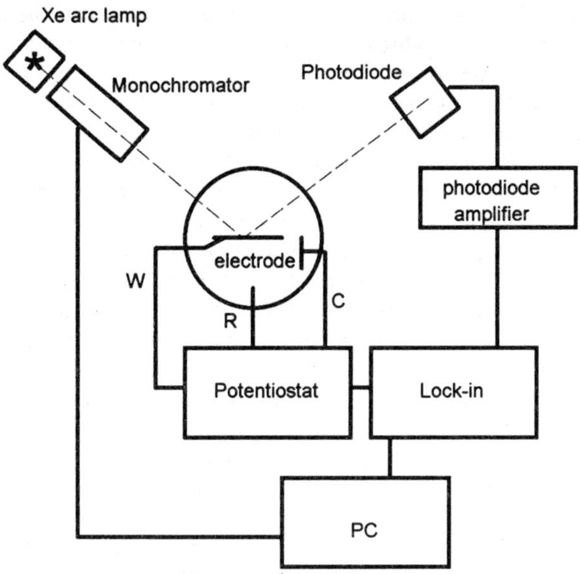

Figure 1 *Experimental set-up for PMR.*

3 RESULTS AND DISCUSSION

3.1 Electrochemical Impedance Spectroscopy

Electrochemical impedance spectroscopy is a powerful technique that allows deconvolution of the different factors contributing to electrode overpotential, and it also provides information about the electrode capacitance. In this work, EIS measurements were made under conditions of chlorine evolution rather than, as is more common, close to equilibrium. As a consequence the information obtained is more relevant to electrode performance under realistic conditions.

In the Tafel region, the current density for chlorine evolution can be expressed as

$$j + \delta j = j_0\, exp\left[\frac{\alpha nF}{RT}(\eta + \delta\eta)\right] \tag{1}$$

Here δj represents the *ac* current perturbation superimposed on the *dc* current density j. The corresponding *ac* perturbation of the overpotential is $\delta\eta$. Under the conditions of the EIS experiment, δj is sufficiently small that the exponential term containing $\delta\eta$ can be linearised. It can be shown that the ratio of the overpotential and current density perturbations defines a Faradaic or charge transfer resistance R_{ct} given by

$$R_{ct} = \frac{\Delta\eta}{\Delta j} = \frac{RT}{\alpha nFj} \tag{2}$$

Measurements of R_{ct} as a function of *dc* current density can therefore be used to obtain the Tafel slope, which can be compared with the value obtained by conventional polarisation measurements.

The advantage of the EIS method is that the ohmic component can be eliminated by frequency response analysis of the impedance. The simplest equivalent circuit consists of R_{ct} in parallel with the double layer capacitance, C_{dl}, and the ohmic resistance is in series with this parallel combination. The impedance response of this combination of elements is expected to be a semicircle displaced along the real axis of the complex plane plot. The high frequency intercept gives the ohmic series resistance, whereas the low frequency intercept is the sum of the ohmic and Faradaic resistances. The maximum of the semicircle occurs at a frequency given by $f_{max} = 1/2\pi R_{ct}C_{dl}$. The analysis therefore gives values for all three parameters.

Figure 2 illustrates the complex plane impedance responses obtained for the RuO_2/TiO_2 coated anode during chlorine evolution in brine at 60° C at *dc* current densities between 40 mA and 140 mA cm^{-2}, and the corresponding Bode plots for the same data are also shown. As predicted by eqn. 2, the diameter of the impedance semicircles decreases with increasing *dc* current density. The frequency of the maximum also varies with current density as predicted by eqn. 2, and ranges typically from 20 -100 Hz. C_{dl} for different samples was found to be independent of current density and to lie in the range 15-30 mF cm^{-2}, in good agreement with the values derived from cyclic voltammetry. The flattening of the semicircles indicates that surface heterogeneity leads to a dispersion in the values of R_{ct} and C_{dl}.

Figure 2 shows that the values of R_{ct} are quite low, and at higher current densities it becomes more difficult to obtain reliable values since bubble evolution introduces noise into the measurements and R_{ct} becomes comparable to the ohmic resistance.

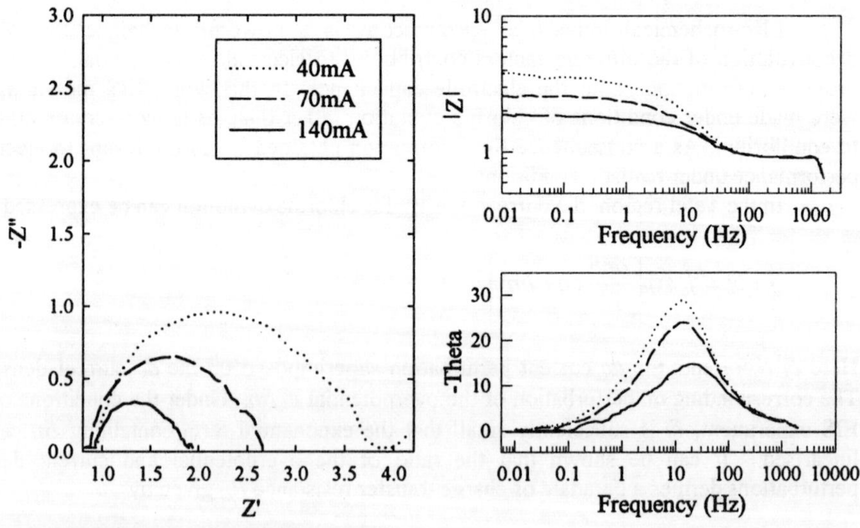

Figure 2 *EIS plots for RuO_2/TiO_2 electrode in chlorine saturated brine at current densities of 40, 70 and 140 mA cm^{-2}. The right hand plot shows the corresponding Bode plots of the same data.*

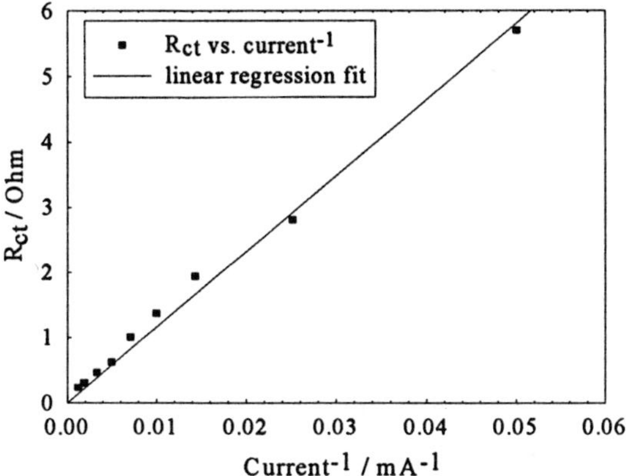

Figure 3 *Plot of R_{ct} vs. 1/j used to determine the Tafel slope(cf. eqn. 2)*

The values of R_{ct} derived from analysis of the EIS response are plotted in Figure 3 according to eqn. 2. The straight line fit of the data to eqn. 2 gives a value of the Tafel slope $dlog_{10}j/d\eta$ = 255 mV/decade. This high value is consistent with a high surface coverage of the electrode with Cl adatoms.

3.2 Open Circuit Potential Decay Measurements

The theory and practice of the OCPD method have been developed by Conway and co-workers[4-6]. The method involves interruption of an initially defined current and observation of the decay of the electrode potential. The initial decay due to the disappearance of the ohmic potential is essentially instantaneous, whereas the subsequent decay takes place over a time scale that extends to seconds. In the case of the chlorine evolution reaction, two reaction mechanisms can be considered.

reaction 1 $$Cl^- \longrightarrow Cl_{ads} + e \quad (3)$$
followed by
reaction 2 $$2\,Cl_{ads} \longrightarrow Cl_2 \quad (4)$$
or alternatively by
reaction 3 $$Cl^- + Cl_{ads} \longrightarrow Cl_2 + e \quad (5)$$

Equations 4 and 5 correspond to chemical and electrochemical desorption of chlorine respectively. If we assume that the transfer coefficients for reactions 3 and 5 are both 0.5, the rates of equations 3 - 5 can be written as

$$v_1 = k_1(1-\theta)exp\left(\frac{F\eta}{2RT}\right) - k_{-1}\theta exp\left(\frac{-F}{2RT}\eta\right) \quad (6)$$

$$v_2 = k_2\theta exp\left(\frac{F\eta}{2RT}\right) - k_{-2}(1-\theta)exp\left(\frac{-F\eta}{2RT}\right) \quad (7)$$

$$v_3 = k_3\theta^2 - k_{-3}(1-\theta)^2 \tag{8}$$

Conway and co-workers[4-6] have shown that the OCPD transients for the two cases should be distinguishable. The theoretical treatment lies outside the scope of the present paper, but the general features of the OCPD curves can be explained by considering the two mechanisms outlined in eqns. 3 - 5. After interruption of the current, the charge stored in the double layer can be consumed by continued discharge of chloride ions, and the overpotential falls. If the equilibrium potential for reaction 1 is more positive than for reaction 3, the overpotential will decay until it passes through the equilibrium potential for reaction 1. Reaction 1 will then reverse its direction as the overpotential falls further, while reaction 3 still proceeds in the forward direction. We now have the situation where reactions 1 and 3 are proceeding in opposite directions. The net reaction that results is the generation of chlorine by desorption without current flow:

$$Cl_{ads} + e \longrightarrow Cl^- \tag{9}$$

$$Cl_{ads} + Cl^- \longrightarrow Cl_2 + e \tag{10}$$

i.e. overall

$$2\ Cl_{ads} \longrightarrow Cl_2 \tag{11}$$

The consequence of this reaction sequence is seen in the overpotential decay as a plateau during which chlorine is desorbed by reactions 9 - 11, and the overpotential decay is temporarily halted because reactions 9 and 10 result in almost zero internal discharge current. The appearance of such a plateau is therefore diagnostic for the reaction scheme involving electrochemical chlorine desorption.

Figure 4 shows a set of OCPD curves plotted on a logarithmic time scale and fitted using the theoretical expressions derived from the kinetic scheme.

The figure shows that data can be obtained over six decades of time, allowing unequivocal fitting to a reaction mechanism. The plateau predicted for the situation described by eqns. 9 - 11 is clearly visible in the transients, so that we can conclude that the chlorine evolution reaction proceeds via electrochemical desorption of the adsorption intermediate. The fit also provides values of the initial overpotential corrected for ohmic drop, and these values can be used to construct Tafel plots.

Closer analysis of the overpotential decay at very short times ($< 10^{-5}$ s) revealed non-idealities attributed to surface heterogeneity and surface roughness. Non-uniform current distributions arising from geometric effects can also be important, particularly at high current densities. It is particularly difficult to identify the ohmic overpotential contribution, and it was concluded that Tafel plots derived by the interrupt method become unreliable at high current densities (in this work up to 1 A cm^{-2}). Nevertheless the OCPD method remains a powerful diagnostic technique that provides a fingerprint of the chlorine reaction over an extended potential and time window. In this respect, the method contrasts with electrochemical impedance spectroscopy which is limited to small perturbations and a smaller effective time (frequency) range.

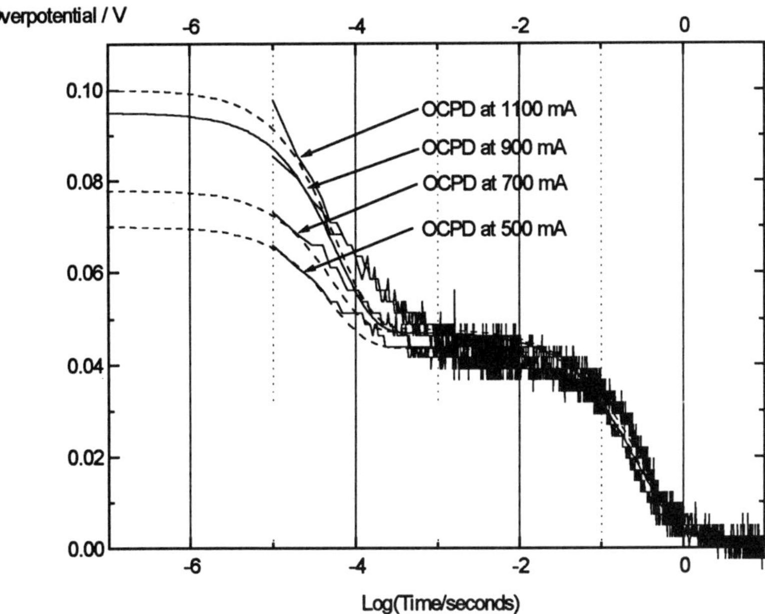

Figure 4 *OCPD transients for RuO_2/TiO_2 electrode in chlorine saturated brine at 60° C for different initial currents. The broken lines illustrate the theoretical fits for two of the transients.*

3.3 Potential Modulated Reflectance Measurements

The reaction schemes outlined in eqns. 3 - 5 do not identify the nature of the adsorbed chlorine intermediate. In the case of RuO_2/TiO_2 electrodes it is reasonable to assume that the redox chemistry of ruthenium plays an important role in the reaction. It is therefore important to develop surface sensitive methods that can be used *in-situ* to study the reactivity and bonding of surface ruthenium. The stability of ruthenium is also of interest, and it is known that bulk ruthenium corrodes rapidly during oxygen evolution. By contrast, RuO_2/TiO_2 electrodes are extraordinarily stable when used as chlorine anodes.

Kötz *et al.*[9] have used differential reflectance spectroscopy to study the formation of RuO_4 during the corrosion of ruthenium during oxygen evolution, and Bewick *et al.*[10] have used *in-situ* infrared spectroscopy to detect soluble Ru(VII) species formed in alkaline solutions. In this work we have used potential modulated reflectance spectroscopy (PMR) to study a model electrode prepared by electroplating ruthenium on a smooth platinum substrate. Figure 5 shows typical cyclic voltammograms for Ru in acid and alkaline solutions. The information provided by the voltammograms is limited. The oxidation and reduction peaks are broad and it is not easy to relate their positions to the standard potentials for different ruthenium oxidation states because they are not reversible. The cyclic voltammograms of RuO_2/TiO_2 electrodes show even fewer features.

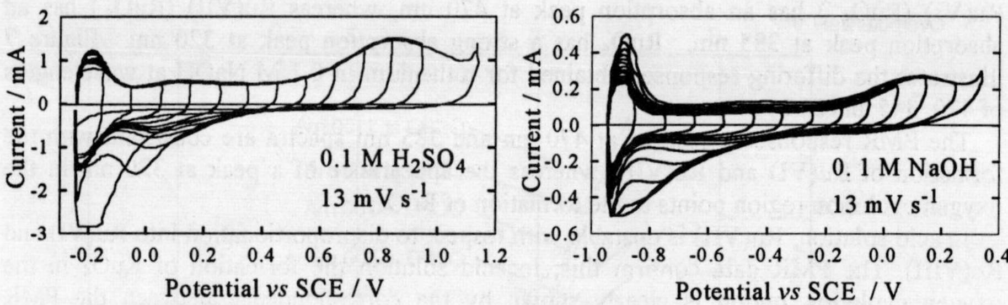

Figure 5 *Cyclic voltammograms for Ru in acid and alkaline solutions. The figures show the effect on the voltammograms of changing the anodic limit.*

PMR is a more powerful technique than voltammetry since it detects the absorption spectra of the different Ru states. In fact, PMR is a form of differential spectroscopy: it detects the difference between the absorption properties of the system at two potentials. It not only provides spectroscopic information but also kinetic information. Kinetic information is obtained from frequency response analysis of the signal and from the phase relationship between the potential modulation and the reflectance response.

An example of a phase sensitive PMR spectrum for Ru in alkaline solution is given in Figure 6. The *dc* potential was set at 0.32 V, and a 100 mV *ac* modulation applied. The in-phase and 90° components of the normalised modulated reflectance response ΔR/R were recorded simultaneously. The change in sign of the response at 420 nm is due to the existence of an isosbestic point in the absorption spectra of the species involved. Detailed comparison with published absorption spectra for different oxidation states of ruthenium showed that the PMR response corresponds to interconversion of RuO_4^{2-} and RuO_4^- [11].

PMR can be used to follow the changes in surface and solution composition as a function of potential. This is done by choosing a fixed wavelength and scanning the electrode potential slowly across the region of interest. The initial oxidation of Ru is expected to form RuO_2. Further oxidation can in principle result in formation of Ru(VI), Ru(VII) and Ru(VIII) species. These can be identified by their absorption spectra[11].

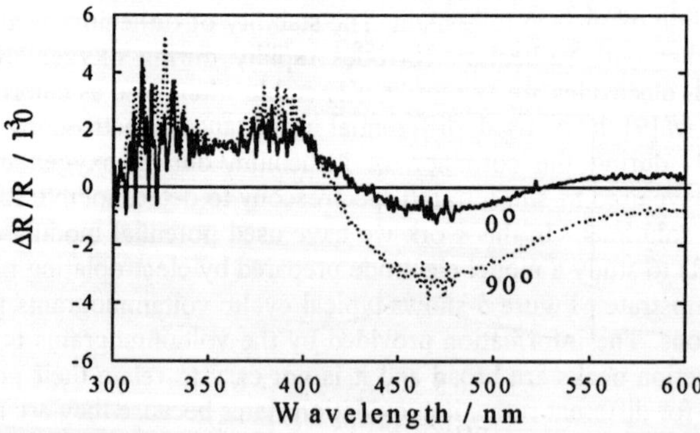

Figure 6 *PMR spectrum showing in-phase and 90° components of normalised reflectance for Ru in 0.1 M NaOH at 0.32 V vs. SCE. Note the crossover at 420 nm corresponding to the isosbestic point for Ru(VI)/Ru(VII).*

Ru(VI) (RuO_4^{2-}) has an absorption peak at 470 nm, whereas Ru(VII) (RuO_4^-) has an absorption peak at 385 nm. RuO_4 has a strong absorption peak at 320 nm. Figure 7 illustrates the differing responses obtained for ruthenium in 0.1 M NaOH at wavelengths of 470, 345 and 320 nm.

The PMR responses measured at 470 nm and 385 nm spectra are consistent with the formation of Ru(VI) and Ru(VII), whereas the appearance of a peak at 320 nm in the oxygen evolution region points to the formation of RuO_4.

In acid solution, Ru(VII) is unstable with respect to disproportionation into Ru(VI) and Ru(VIII). The PMR data confirm this; in acid solution the formation of RuO_4 in the oxygen evolution region is clearly shown by the correspondence between the PMR spectrum (Figure 8) and the absorption spectrum of RuO_4 [11].

The PMR results show that it is possible to identify redox reactions taking place at ruthenium electrodes. Work is now in progress to extend the method to the study of chlorine evolution at Ru and RuO_2/TiO_2 electrodes.

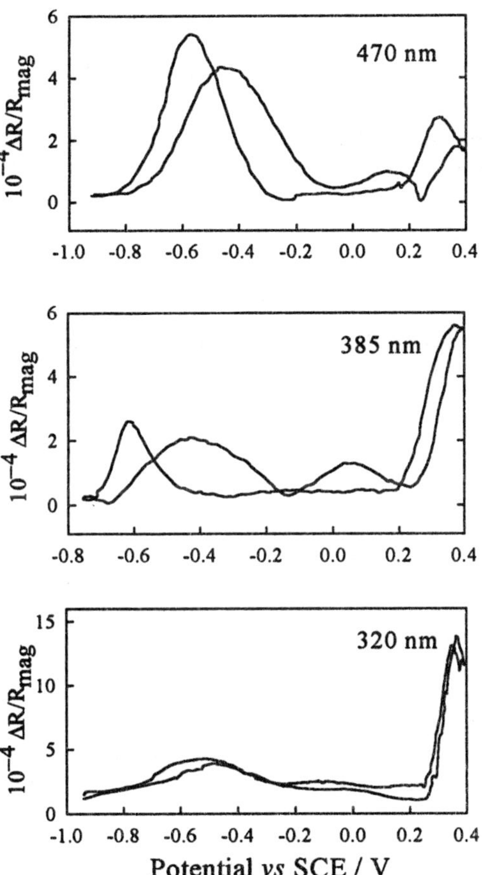

Figure 7 *PMR response as a function of potential for Ru in 0.1 M NaOH at the wavelengths shown. Note that the plots show the magnitude of ΔR/R.*

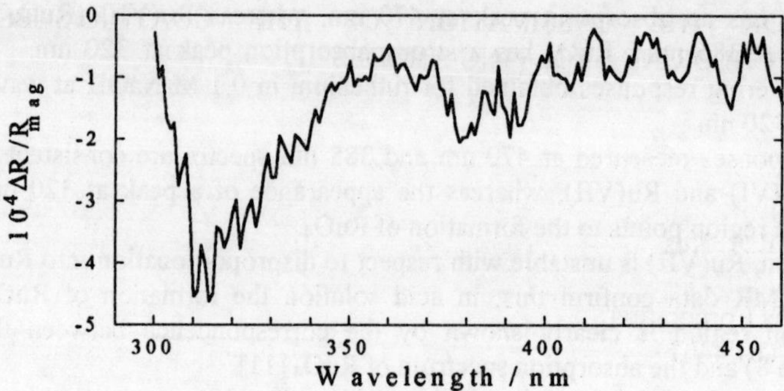

Figure 8. *PMR response for Ru in 0.1 M H_2SO_4 at 1.15 V vs. SCE showing peaks at 320 nm and 385 nm corresponding to RuO_4.*

4 CONCLUSIONS

Further progress towards understanding the mechanism of catalysis of the chlorine evolution reaction requires application of a range of experimental techniques. Electrochemical methods give little information about the chemical composition of the surface or of the reaction intermediates. The use of *in-situ* spectroscopies such as PMR offers a way forward toward the ultimate objective of understanding catalysis and stability at the molecular level.

Acknowledgements

Anthony Shingleton acknowledges support from ICI. Rachel Walker acknowledges EPSRC and ICI. The authors also thank Professor Brian Conway for useful discussions.

References

1. C. Dodd, D.R. Hodgson, L.M. Peter and A. Shingleton in *Modern Chlor-Alkali Technology*. Vol. 6. pp. 98-109. Ed. R.W. Curry, Roy. Soc. Chem., 1995.
2. B.E. Conway and G. Ping. *J. Chem. Soc. Faraday Trans.*, 1990, **86** 923.
3. B.E. Conway and L. Bai, *J. electrochem. Soc.*, 1991, **138**, 2897.
4. D.A. Harrington and B. E. Conway, *J. electroanal. Chem.*, 1987, **221**, 1.
5. B.E. Conway, L. Bai and L. Gao, *J. Chem. Soc. Faraday Trans.*, 1990, **89**, 91.
6. B.E. Conway, P. Gu, L. Bai, L. Gao and R. Brousseau, *Electrochim. Acta*, 1992, **37**, 2145.
7. L.M. Peter, *Phil. Trans. Roy. Soc*. Ser. A, 1996, **354**, 1613.
8. A. Shingleton, Ph.D. Thesis, University of Bath, 1996.
9. R. Kötz, S. Stucki, D. Searson and D.M. Kolb, *J. electroanal. Chem.*, 1984, **177**, 211.
10. A. Bewick, C. Gutierrez and G. Larramona, *J. electroanal. Chem.*, 1992, **332**, 155.
11. R.E. Connick and C.B. Hurley, *J. Am. Chem. Soc.*, 1956, **74**, 5012.

7

EVIDENCE FOR THE PASSIVATION OF THE COATING/SUBSTRATE INTERFACE IN CHLORINE EVOLVING ANODES

K.L. Hardee and R.A. Kus

ELTECH Systems Corporation
Fairport Harbor, OH 44077
USA

1 INTRODUCTION

Dimensionally Stable Anodes (DSA®)* consist of titanium metal substrates with a surface coating consisting principally of platinum group metal oxides and valve metal oxides as introduced by H. Beer in the 1960's.[1-2] These have become the standard anodes for chlor-alkali, chlorate and hypochlorite systems where the platinum group metal oxide is ruthenium oxide (RuO_2) or iridium oxide (IrO_2). These anodes have demonstrated long lifetimes in various types of chlorine cells, including diaphragm cells where lifetimes in excess of 15 years have been observed. Understanding the failure mode in such long lived anodes is important in order to develop strategies for timely replacements before widespread failures occur that would be detrimental to a plant's operation.

The key characteristic of the anode coating is the maintenance of a low operating potential over the course of the life of the anode. Ultimately, though, the anode voltage begins to rise due to either the presence of surface deposits or due to passivation of the coating/substrate interface, where passivation is the growth of a non-conductive titanium oxide layer between the coating and the substrate. Surface deposits not only affect the anode voltage directly but can lead to passivation of the anode, indicated by an exponential rise in voltage with time. In many cases, if caught early, surface deposits can be removed from the anode and the voltage restored. However, once passivation is initiated, the anodes will need to be reprocessed to remove the coating and oxide layer and a new coating applied in order to be restored ("recoating").

There are three main categories of anode failures: a) loss of coating b) blinding deposits and c) passivation of the coating/substrate interface. Loss of coating, e.g. wear, is a normal phenomenon that occurs gradually over the course of the operation of an anode. Both chemical/electrochemical and mechanical forces contribute to the slow attrition of the coating. When the coating becomes thin enough the anode voltage will rise as the substrate passivates. Likewise, certain chemical agents (e.g. fluoride) can lead to rapid loss of the coating and failure of the anode (i.e. high voltages).

Deposits are generally the result of impurities in the electrolyte. Even ppm quantities can provide significant deposits over a long period of operation. Higher quantities can

* DSA is a registered trademark of ELTECH Systems Corporation

lead to a rapid deposit formation and, again, a high anode potential. Failure to remove the deposits can lead to damage to the anode, e.g. passivation, such that cleaning will not restore the potential.

Finally, the coating may fail from the growth of a non-conductive titanium oxide layer between the coating and the substrate (i.e. passivation). The growth rate of this layer correlates with the amount of oxygen evolved at the anode surface. For chlorine anodes with low oxygen evolution, lifetimes can be measured in years, but under 100% oxygen evolution the lifetime is often a matter of hours or days. Surface deposits can also accelerate the formation of this layer by inhibiting access of chloride to the anode surface which results in a higher potential and then passivation of the substrate. Passivation of the substrate can occur with relatively high amounts of coating still present. In order to detect the presence of layers between the coating and the substrate in a non-destructive manner, and at a thickness not easily observed in cross-sections, an Electrochemical Impedance Spectroscopy (EIS) technique has been developed. Surface science techniques have also been utilized to examine possible changes in the coating composition. These techniques have been used to examine anodes used in chlorine production.

The actual growth of a non-conductive titanium oxide layer between the substrate and the coating has been observed in oxygen evolution applications where this is the normal failure mode. However, in chlorine systems the wear mechanism (slow attrition of the coating) has been the "usual" failure mode for anodes. The coating gradually thins over the course of operation until a point is reached where recoating is deemed advisable to prevent massive failures. However, with long operating times or with the presence of significant surface deposits, the passivation mechanism has been found to also occur.

In operation a DSA anode typically has a low operating potential over most of its effective lifetime. However, when passivation of the anode occurs the voltage rises exponentially. A typical voltage vs. time curve is shown in Figure 1. This rise in voltage is believed to be due to passivation of the interface between the coating and the titanium

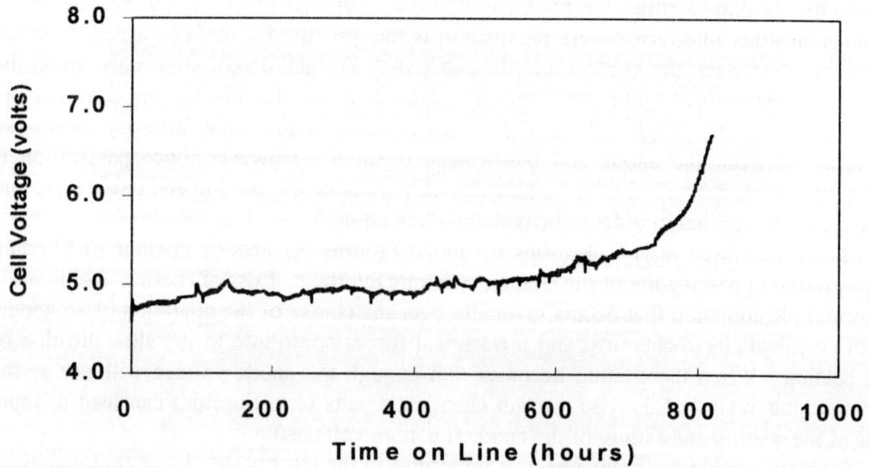

Figure 1 *Voltage vs. Time Indicating Passivation of Anode Operated in Laboratory Test Cell with Sulfate Electrolyte*

substrate. However, higher resistances due to changes in the coating itself or to the presence of surface deposits can also produce elevated voltages, although these may not be exponential with time. Distinguishing between these effects as early as possible is important in commercial operation to be able to decide whether a simple cleaning of the anode is sufficient or whether passivation has begun which necessitates recoating of the anodes.

Since anodically formed titanium oxide is essentially an insulator, the presence of very thin layers is sufficient to increase the overall resistance of the anode and thus the cell voltage. Due to the sensitivity of chlorine production to power costs, small increases in anode voltage require remedial action. Often anodes are removed from service when the potential has increased only a few hundred millivolts. Direct confirmation of the presence of the passivation layer, using Scanning Electron Microscopy (SEM) for example, under these operating conditions is•extremely difficult.

However, in oxygen evolving systems the anode voltage occasionally will exceed 10 volts before the anode can be removed. The SEM photograph of a cross-section of such an anode is shown in Figure 2. The applied coating, in this case, contained Ta and Ir

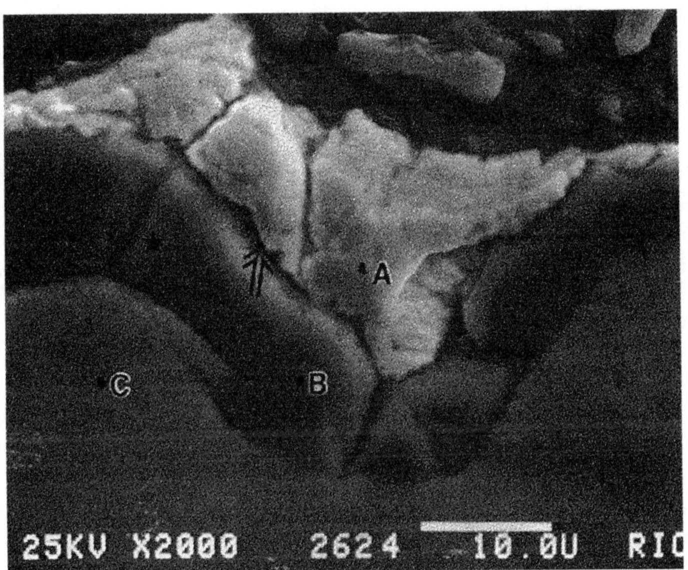

Figure 2 *SEM Cross-Sectional View of Passivated Anode*
A: Coating B: TiO_2 Layer C: Titanium Substrate

oxides. The layer between the substrate and the coating was clearly identified by Energy Dispersive X-Ray Spectrometry (EDS) as being composed of both titanium and oxygen, consistent with TiO_2. Similar results have been shown by others.[3] Thus the growth of a titanium oxide layer between the coating and the substrate is possible.

Another alternative explanation of a voltage rise with used anodes might be depletion of the surface concentration of the electrocatalyst, e.g. ruthenium. Alterations in the bulk Ti:Ru ratio can increase the anode potential.[4] However, simple measurements of the surface concentration of Ru can be misleading, as the initial surface concentration is significantly less than that of the bulk concentration.[5] A depleted layer would also have to be relatively uniform over the surface to account for the high Single Electrode Potential (SEP). XPS and Auger spectroscopy measurements of the surface concentrations of the coating components (e.g. Ti and Ru) for various anodes (from cells for both oxygen and chlorine evolution) show wide variations , although anode potentials are near normal.

2 EXPERIMENTAL

Electrochemical Impedance Spectroscopy (EIS), or AC Impedance, measurements were done with a Princeton Applied Research (PAR) M273 potentiostat/galvanostat with a Solartron 1250 Frequency Response Analyzer (FRA). Data acquisition was done using in-house software. A standard three-electrode glass cell was used with a DSA counter electrode and a Saturated Calomel Reference Electrode (SCE). The reference probe was a glass pipette fitted through the counter electrode. A Pt wire was fitted through the probe tip to minimize stray capacitance. For the on-line EIS measurements during lifetesting, a PAR model 173 potentiostat and a Solartron 1255 FRA were used with switching relays which disconnected the rectifier under computer control. There was a 5 minute delay prior to starting the actual EIS measurement. The frequency sweep was from 65000 Hz to 0.1 Hz with an AC amplitude of 0.1 mA/cm^2 superimposed on a DC current of 1 mA/cm^2.

Single Electrode Potential (SEP) measurements for chlorine evolution were done in 300 gpl NaCl, pH 1.0, at 75 °C. Anode potentials are reported vs. SCE and are inclusive of an iR term since contributions from the electrolyte, deposits or passivation cannot be readily separated. The potentials were measured with the reference probe tip fixed at 4.0 mm from the anode surface to provide relative comparisons of the various samples.

Lifetests were done in glass cells with Zr cathodes using an electrolyte of 285 gpl Na_2SO_4 + 60 gpl $MgSO_4$ at pH 2.0 and ~65 °C with a current density of 15 kA/m^2.

X-ray fluorescence measurements were done with a TEFA Monitor (HNU Systems) using either a Mo or W X-ray tube source.

Surface science analytical techniques used were X-Ray Photoelectron Spectroscopy (XPS) and Auger spectroscopy. The XPS measurements were taken using a Physical Electronics Model 5600 Instrument. X-ray source is monochromatic Al K_α X-rays. For depth profiling, argon ion sputter rates were ~45 Å per minute. Analysis area is approximately 800 μm (0.8 mm) diameter, with a depth of ~30-50 Å.

The Auger measurements were taken using a Physical Electronics Model 660 Scanning Auger Microscope. Analysis area varies depending on magnification of sample with typical magnifications of 500-1000x. Analysis depth is ~30-50 Å using source beam energy of 10 keV. Measured survey kinetic energy range is 0-2000 eV. For depth profiling, argon ion sputter rates are variable, typically 200-300 Å per minute.

3 RESULTS AND DISCUSSION

EIS involves the application of a small AC signal to an anode and monitoring the response which in the present application is the phase shift of the signal as a function of frequency. A typical impedance spectra of an unused anode with a RuO_2 coating is shown in Figure 3. The peak in the spectrum at low frequency is principally due to the coating/electrolyte interface. An anode can be operated in a sulfate electrolyte until it passivates, i.e. the voltage vs. time curve appears as in Figure 1. A sample of a standard DSA anode was operated in the laboratory lifetest system and the EIS spectra were taken periodically during this lifetest utilizing the computer controlled, on-line system. These curves appear as in Figure 4. An additional peak in the phase spectra in the 1000-3000 Hz region can be seen to grow with time. This peak appears to be associated with the growth of the titanium oxide layer between the coating and the substrate.

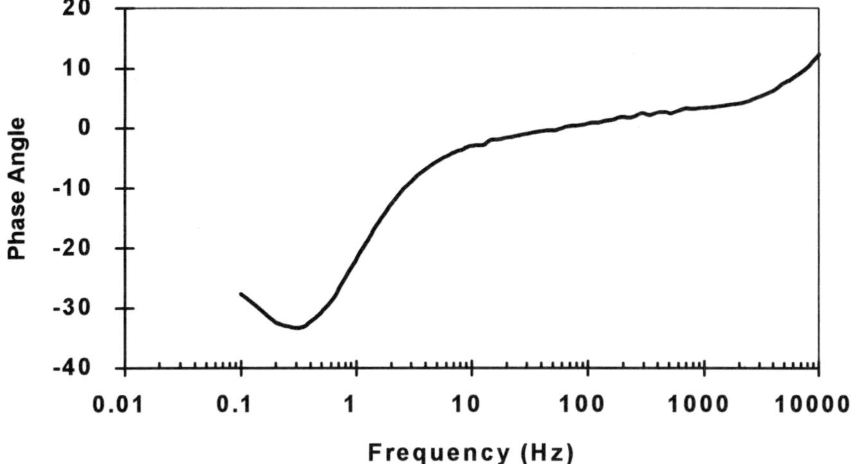

Figure 3 *Initial EIS Spectrum of DSA Anode*

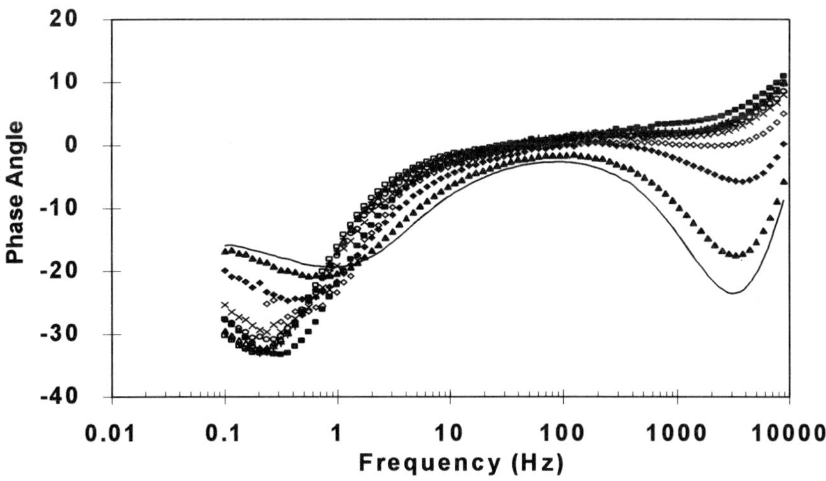

Figure 4 *EIS Spectra with Time on Line in Laboratory System with Sulfate Electrolyte Lifetime:* ■ *0%* + *11%* △ *22%* □ *28%* ○ *58%* x *72%* ✧ *85%* ◆ *91%* ⅄ *97%* **-***100%*

Initial plots of unoperated anodes are shown in Figure 5 for an anode coating containing either $RuO_2/IrO_2/TiO_2$ or RuO_2/TiO_2 only. Similar plots have been generated for coatings with different valve metal components, e.g. Ti, Ta or Sn, and different ratios of Ru and Ti. Thus the composition of the coating does not significantly affect the spectra. Plotting the peak height for the data in Figure 4 vs. the cell voltage shows a good correlation as seen in Figure 6. Thus the rise of this phase shift peak appears directly related to the increase in anode potential.

To rule out other possible changes in the coating that might contribute to a change in the EIS spectra, additional experiments were done. These were designed to demonstrate that the higher frequency phase angle peak was due to substrate passivation and not compositional changes in the coating itself.

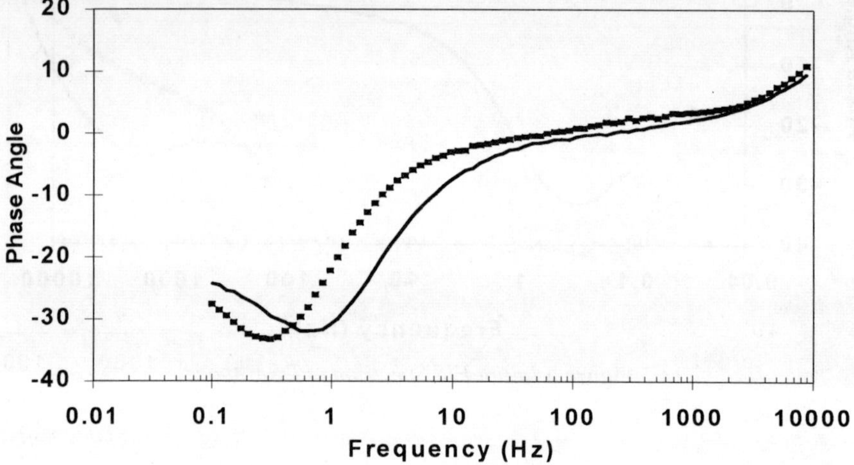

Figure 5 *Initial EIS Spectra of RuO_2/TiO_2 (▬) and $RuO_2/IrO_2/TiO_2$ (■) Coatings*

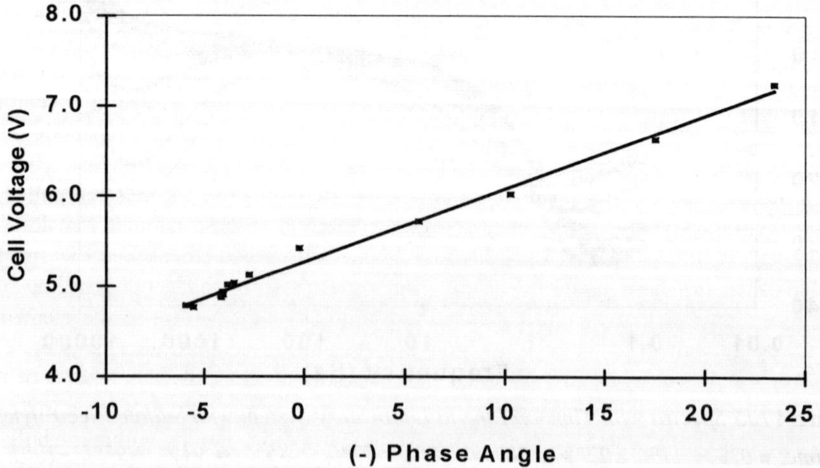

Figure 6 *Correlation of Cell Voltage with Change in Phase Angle for Data in Figure 4*

First, a sample was prepared with a coating consisting only of IrO_2 (i.e. no valve metal oxide was added to the matrix). This sample was operated in the laboratory lifetest system under oxygen evolving conditions to failure (high voltage). After the anode failed, X-ray fluorescence measurements indicated that more than 50% of the original coating was still in place. EIS measurements again indicated the development of a peak in the higher frequency region as the anode voltage increased, similar to that for coatings consisting of RuO_2/TiO_2. The initial and final EIS spectra are shown in Figure 7. Since the coating consisted only of IrO_2, no compositional change of the coating was possible.

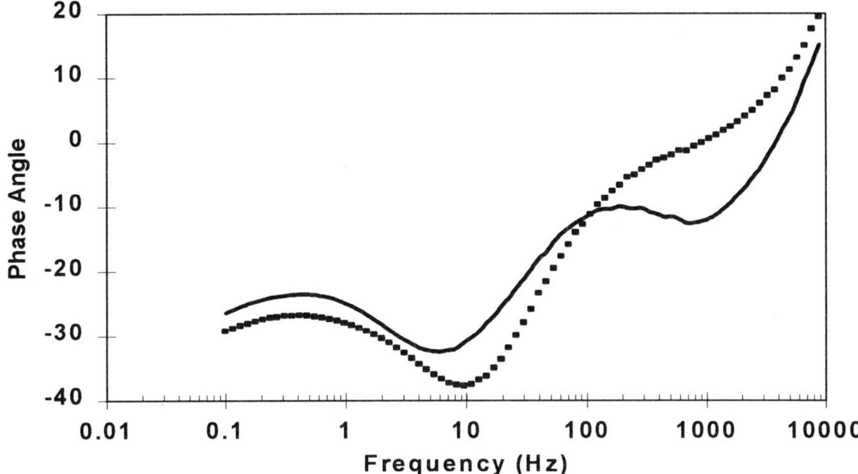

Figure 7 *EIS Spectra for IrO_2-only Coating Before and After Passivation in Sulfate Electrolyte (■ Initial ▬ Passivated)*

Where changes in composition are possible, e.g. in a RuO_2/TiO_2 coating, the question of the effect of changes in the ratio of the coating components arises. To demonstrate the effect of an increased TiO_2 content at the outer surface, a solution of tetra-butylorthotitanate was applied to the surface of a standard RuO_2/TiO_2 coating, followed by high temperature baking to form several layers of TiO_2 on the surface of a coating. The EIS spectra (Figure 8) show essentially no change even though the anode SEP increased by 500 mV with the higher amounts of TiO_2.

A series of laboratory coupons were operated in the laboratory test cell and pulled at intervals along the voltage vs. time curve. Surface analysis of these samples was done to determine the Ti:Ru ratio in the outer surface. Anode SEP measurements and EIS spectra were also taken for these samples. Using Auger spectroscopy and depth profiling, the Ti:Ru ratios were calculated from the measured atomic weight percent of the elements. The profiles are shown in Figure 9 for a control, partially used and failed sample. The failed sample had a chlorine SEP value of nearly 3 volts vs. SCE, well in excess of the potential where anodes are usually removed from service. While there is a large increase in the Ti:Ru ratio at the outer layers (ca. 50 nm), the ratio drops rapidly within the bulk of the coating. A comparison of the near surface ratios (from XPS data) of several of the samples (Figure 10) and the EIS phase shift peak, as a function of the anode potential, finds that while the phase shift peak tracks the potential well, there is no clear correlation between the surface ratios and the potential. This is consistent with the increased anode potential being due to the growth of the TiO_2 layer at the coating/substrate interface and not simply surface changes in the coating. The EIS technique, then, can be a powerful tool to detect the presence of this layer in operated anodes.

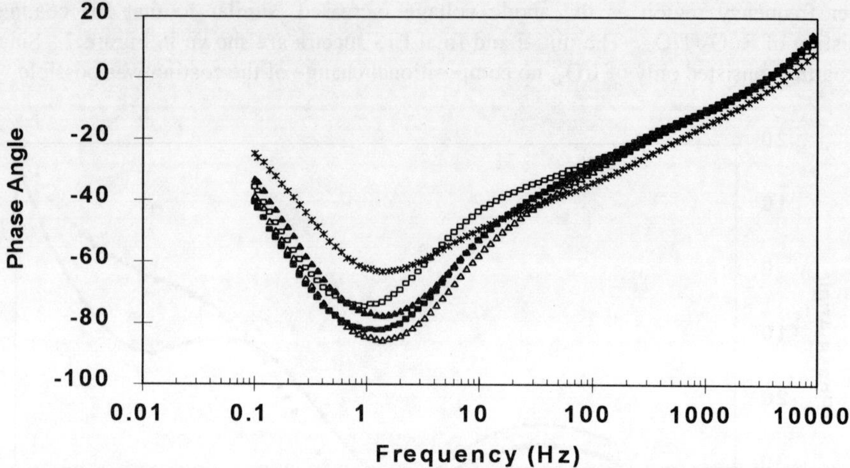

Figure 8 *EIS Spectra of RuO_2 /TiO_2 Coating with TiO_2 Layers Applied to the Surface (Number of Layers Applied: □ 0 ■ 6 ✧ 10 ● 14 X 26)*

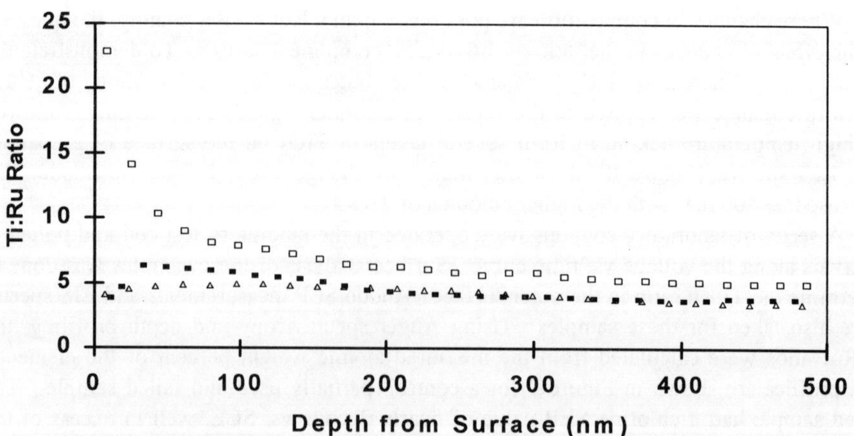

Figure 9 *Auger Depth Profile of Laboratory Tested Anodes Showing Changes in Ti:Ru Ratio [Anode Potential: □ 2.85 Δ 1.27 ■ 1.30 (V vs. SCE @ 3 kA/m^2)]*

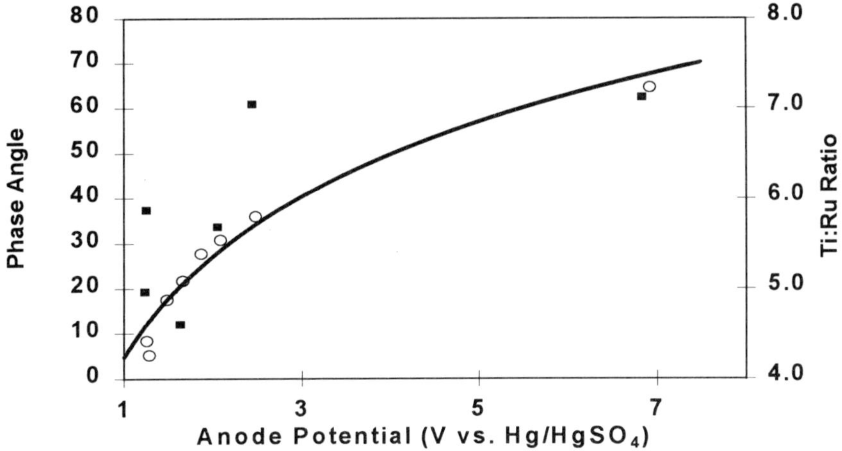

Figure 10 *Correlation of EIS Phase Angle and Ti:Ru Ratios with Anode SEP (■ Ti:Ru ○ Phase)*

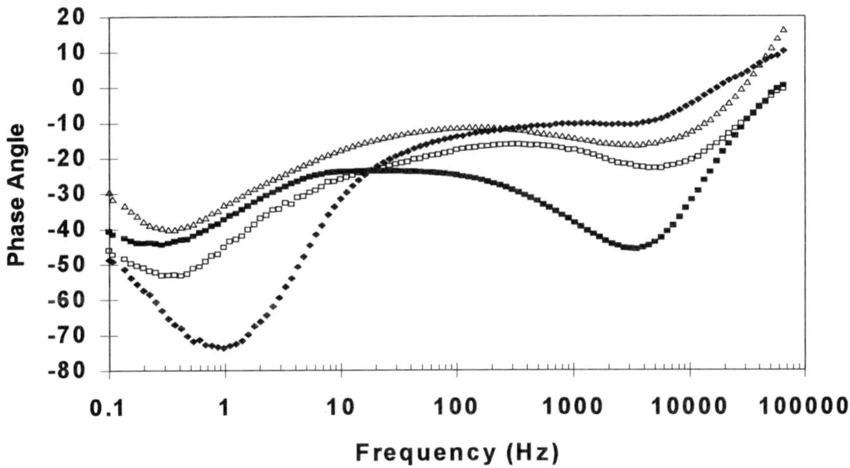

Figure 11 *EIS Spectra of Used Commercial Diaphragm Anodes Indicating Passivation Anode Potentials in NaCl at 3 kA/m^2: ◆ 1.3 Δ 2.0 □ 2.1 ■ 2.5 (V vs. SCE)*

A number of commercial diaphragm anode samples have been examined with both EIS and with Auger to determine whether passivation is occurring. Figure 11 contains the EIS spectra of several samples from anodes from one diaphragm cell indicating a range of phase shifts, corresponding to varying potentials. The appearance of the higher frequency phase shift indicates the onset of passivation with these anodes, requiring recoating to restore performance. Auger depth profiling, though, again shows no clear correlation, as the Ti:Ru ratios are not significantly different from a sample with a normal SEP and a normal EIS spectrum. In fact, Figure 12 shows the depth profile of a 15 year-old diaphragm cell anode with a normal SEP (< 1.2 V vs. SCE) but with higher Ti:Ru ratios. Thus while changes in the Ti:Ru ratio of the coating can occur, their direct influence on anode potential appears minimal. Passivation of the substrate is the more important effect.

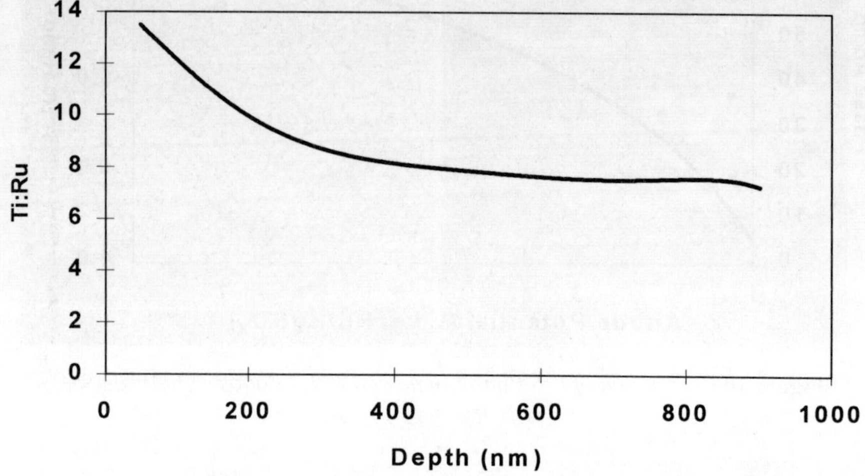

Figure 12 *Ti:Ru Ratios by Auger Depth Profiling for Used (15 yrs) Commercial Diaphragm Anode with Good Anode Potential (1.17 V vs. SCE at 9 kA/m^2)*

Commercially operated anodes often have surface deposits, such as iron oxide on diaphragm cell anodes, which will also affect the anode potential. Often, though, these deposits can be removed and the anode potential restored to near normal. SEM photographs of typical deposits are shown in Figure 13 before and after the deposit is removed from the surface. With the deposit present the SEP was 1.30 V vs. SCE at 8.8 kA/m^2. After cleaning the anode the potential was restored to a normal value of 1.15 V. EIS spectra of such anodes (Figure 14) indicate no passivation peak, either before or after removal of the deposit. Also there was essentially no change in the Ti:Ru ratios (Figure 15) with cleaning, again indicating that the higher ratios, compared with a new anode, do not significantly influence the anode potential. In this case the mere presence of the deposit was sufficient to raise the anode voltage. However, if surface deposits are not removed soon enough, they can accelerate the onset of passivation.

4 CONCLUSIONS

Understanding the origin of high voltages in commercial anodes is important for determining how to recover the anode performance. Surface deposits can often be removed to restore the voltage, but an anode with a passivation layer needs to be reprocessed (recoated). The use of impedance techniques (EIS) has been shown to be a valuable tool for distinguishing between the higher voltages due to surface deposits and substrate passivation. Surface analysis results along with anode potential and EIS data indicates that depletion of surface concentrations of Ru is not a major contributor to elevated anode potentials.

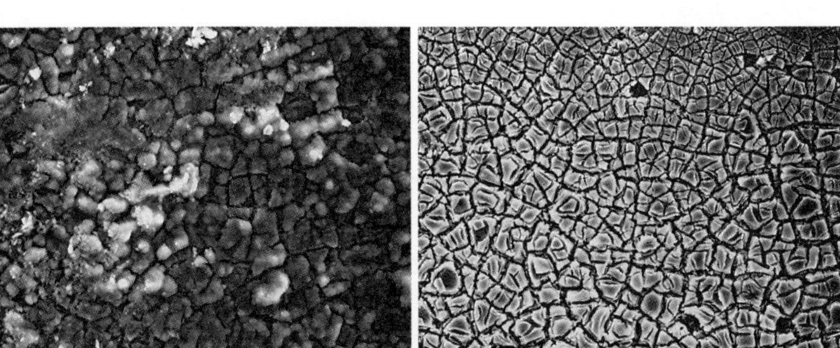

Figure 13 *SEM View of Commercial Anodes Showing Removal of Surface Deposit*

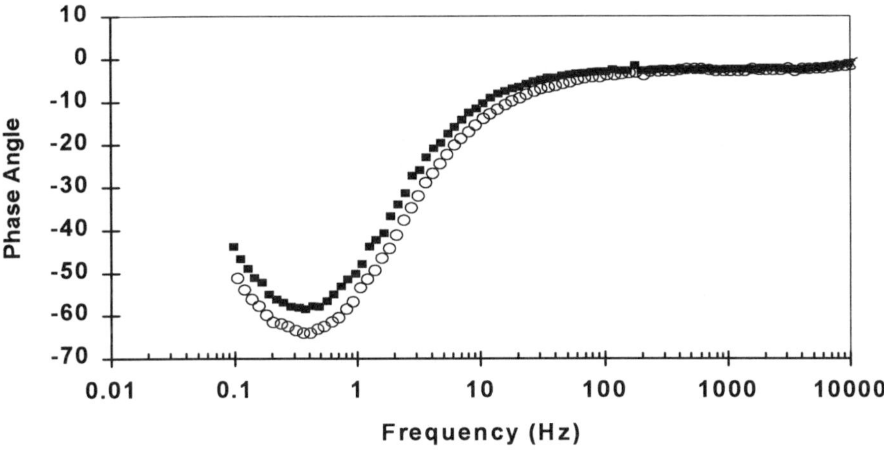

Figure 14 *EIS Spectra for Commercial Anodes with and without Surface Deposit Anode Potentials (V vs. SCE): ■ 1.30 (As Received) □ 1.15 (After Cleaning)*

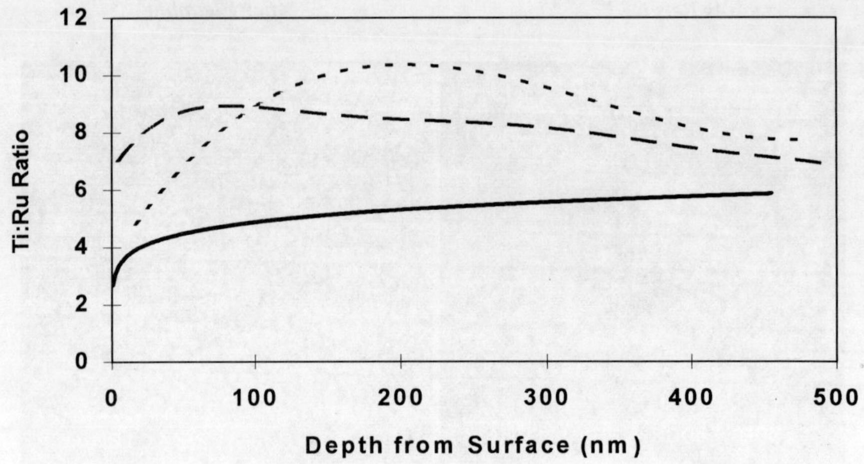

Figure 15 *Auger Depth Profile for Commercial Anodes with and without Surface Deposit (--- Cleaned —— With Deposit —— New Anode)*

5 REFERENCES

1. H. Beer, British Patent 1,147,442, 1969.
2. H. Beer, US Patent 3,632,498, 1972.
3. S. Katowski and B. Busse, "Proceedings of the Symposium on the Performance of Electrodes for Industrial Electrochemical Processes", The Electrochemical Society, Pennington, NH, 1989, p. 245-258.
4. J. Aromaa, Dissertation, Helsinki Univ. of Technology, Report TKK-V-A12, 1994.
5. A. De Battisti, G. Lodi, M. Cappadonia, G. Battaglin, R. Kotz, *J. Electrochem. Soc.*, 1989, **136(9)**, p. 2596-2598.

8

CHLORINE BUBBLE NUCLEATION, GROWTH AND DETACHMENT: THE INFLUENCE OF ANODE STRUCTURE

Sarah L. Gilbert, Lorraine Harris, Matthew W. Carr and A. Robert Hillman
Department of Chemistry, University of Leicester
Leicester LE1 7RH

David R. Hodgson
ICI Chemicals and Polymers Ltd
ETB Technical Centre, P.O. Box 9 Castner Kellner
Runcorn, Cheshire WA7 4JE

1 INTRODUCTION

In any industrial process, product separation is a key technical and economic issue. In the chlor-alkali process, chlorine separation is conveniently accomplished via a phase change: the evolution of chlorine bubbles. Here we describe measurements of this key step.

1.1 Background

Most studies of chlorine electrogeneration from chloride fall into one of two categories. At one extreme, the fundamentals of electron transfer during chloride oxidation have been the subject of a number of theoretical[1] and experimental[2,3,4,5,6] studies. In particular, the roles of adsorbed chlorine atoms and the contributions of "$Cl^{\bullet}_{ads} + Cl^{\bullet}_{ads}$" and "$Cl^{\bullet}_{ads} + Cl^{-}$" steps as chlorine generating reactions have received considerable attention. At the other extreme, the engineering implications of bubble swarms for membrane technology[7] and cell design[8,9] have been considered in detail.

Between these two lies the evolution of chlorine bubbles at the anode. This is poorly understood and, by inference, may be far from optimised. There is no doubt that mixed RuO_2-TiO_2 (RTO) anodes have dramatically improved anode performance in terms of the electron transfer and chlorine generating steps (see above). So, now that kinetic limitations associated with these steps are largely eliminated, we focus on a subsequent possible kinetic limitation: the evolution of chlorine bubbles from chlorine dissolved in solution.

1.2 Key Issues

There are several aspects of the bubble evolution process that merit attention. First, at high current density, when the chloride oxidation rate exceeds the mass transport rate of dissolved chlorine away from the electrode/solution interface, there will be local supersaturation of the solution with dissolved chlorine. The level of supersaturation required to evolve chlorine gas bubbles is determined both by the chemical composition and physical structure of the anode; the former clearly has catalytic implications.

Second, there is the question of whether bubble evolution is homogeneous (within the supersaturated solution) or heterogeneous (at the anode/solution interface). In both cases, the number and size of bubbles have implications for the contribution to the operating potential of the ohmic potential drop.[10] In the latter case, the presence of a bubble at the interface necessarily obscures the electrode surface, making it unavailable for chloride oxidation. The extent of this obscured area is determined by interfacial energetics, characterized by surface tension and contact angle.

1.3 Objectives

Heterogeneous bubble evolution, which we shall show is the predominant pathway, involves three stages: nucleation, growth and detachment. Our first objective is to deduce which of these is rate limiting. Having done this, our second objective is to explore the dependence of the relevant process on anode characteristics, represented (a) chemically by Ru:Ti ratio in the RTO coating, (b) physically by surface morphology, and (c) through operating conditions by chlorine supersaturation and pressure. In the longer term we will use this knowledge to inform anode design.

2 THEORY

2.1 Nucleation, Growth and Detachment

The overall bubble evolution process involves nucleation, growth and (in the heterogeneous case) detachment. Various parameters are known to influence the rate of each process: heterogeneous nucleation is a function of the surface roughness,[11] growth is a function of the rate of diffusion of gas into the bubble,[12] and detachment radius is determined by a balance of surface tension and detachment forces.[13] Nucleation is a non-equilibrium process in which a new phase is formed from an existing phase. In the present context, it is the aggregation of chlorine molecules in solution to form a critical-sized cluster which can then grow spontaneously. We now consider the thermodynamics and kinetics of this process and the way in which it is influenced by the presence of a solid surface, here the anode.

2.2 Thermodynamics of Heterogeneous Nucleation

There have been several significant contributions to the thermodynamics of heterogeneous nucleation[14,15,16], reviewed in detail elsewhere[17,18,19]. We therefore restrict our description to a summary of the key points relevant to the case of interest: a liquid supersaturated with dissolved gas molecules.

Bubble formation will occur spontaneously because the nucleating species has a higher chemical potential (μ) in solution than in the gas phase. At any temperature (T) and pressure (P), the supersaturation (σ) describes the concentration of dissolved gas in solution (c_2) with respect to its equilibrium concentration over a flat surface (c_{2e}) and is defined by

$$\sigma = (c_2 / c_{2e}) - 1 \qquad (1)$$

Upon creation of the new phase, the change in Gibbs energy (ΔG_v) per unit volume (v) of the new phase is given by $\Delta\mu/v$. For spherical bubbles of radius r, the bulk contribution (ΔG_b) to the Gibbs energy change is then $-(4/3)\pi r^3 \Delta G_v$. Against this, there is a positive contribution to the Gibbs energy, associated with the work required to form the interface between the daughter phase (bubble) and the mother phase (solution). For a spherical bubble, this surface term (ΔG_s) is given by $4\pi r^2 \gamma$, where γ is the surface tension. The total Gibbs energy change is then

$$\Delta G = 4\pi r^2 (\gamma - \tfrac{1}{3} r \Delta G_v) \tag{2}$$

For small clusters of chlorine molecules (small r), the surface term dominates and Gibbs energy *increases* with r; bubbles then have a tendency to re-dissolve. For large clusters of chlorine molecules (large r), the bulk term dominates and Gibbs energy *decreases* with r; bubbles can then grow freely. Thus a bubble must achieve a critical size, r*, in order to grow, as illustrated in Figure 1. r* is defined by the turning point in ΔG(r) and is given by $2\gamma/\Delta G_v$.

2.3 Thermodynamics of Homogeneous Nucleation

As discussed elsewhere[15,20] homogeneous nucleation requires much higher values of supersaturation than does heterogeneous nucleation. In the presence of a solid surface, heterogeneous nucleation will therefore restrict the supersaturation to values below that at which homogeneous nucleation becomes detectable. In the particular case of electrochemical generation of the nucleating species, the region of highest supersaturation is necessarily at the solid/liquid interface, further favouring heterogeneous nucleation. We can therefore neglect homogeneous nucleation.

2.4 Nucleation Kinetics

Nucleation occurs as the result of random fluctuations in the system. The rate of nucleation (J) can be described by an Arrhenius-type expression

$$J = Z \exp(-\Delta G^* / kT) \tag{3}$$

where Z is a constant and ΔG* is the Gibbs energy required to generate a critical cluster.

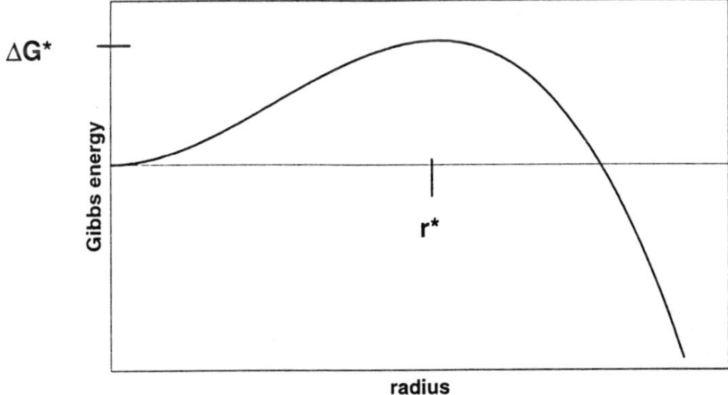

Figure 1. *Schematic diagram showing the Gibbs energy change as function of bubble radius. r* is the critical radius.*

Extending the thermodynamic treatment summarized above, Wilt[21] and Lubetkin[22] were able to express equation (3) in terms of measurable and calculable quantities. The analysis was particularized to the case of nucleation sites modelled as conical pits of half-angle β. The nucleation rate (J) is calculated to be

$$J = N^{2/3} f_{3c(\theta,\beta)} \left(\frac{2\sigma B}{\pi m f_{1c(\theta,\beta)}} \right)^{1/2} \exp\left(\frac{-16\pi\gamma^3 f_{1c(\theta,\beta)}}{3kT(\sigma P)^2} \right) \quad (4)$$

where $N^{2/3}$ is the surface density of nucleation sites, m is the molecular mass, and $f_{3c(\theta,\beta)}$ and $f_{1c(\theta,\beta)}$ are calculable trigonometric functions that describe the nucleation site geometry. Equations (3) and (4) are clearly of the same form, with ΔG* equal to $-16\pi\gamma^3 f_{3c(\theta,\beta)}/3(\sigma P)^2$.

The nature of equation (4) shows the importance of (a) interfacial morphology (via the $f_{(\theta,\beta)}$ functions), (b) interfacial energetics (via γ), and (c) operating conditions (via σ and P). The effects of surface roughness have been studied previously[23] and previous work on these issues has been reviewed elsewhere.[17,18,19] Here we explore the influence of σ and P.

3 EXPERIMENTAL

3.1 Pressure Release Bubble Nucleation (PRBN) Instrumentation

The apparatus is a modification of a design described elsewhere.[24] It comprises a PTFE-lined stainless steel vessel inside which is placed an annealed glass beaker containing a disc fabricated from the nucleating surface under study. The RTO samples were 4 mm discs prepared by coating a roughened titanium substrate in Ru and Ti precursors, which were then thermally oxidised. The sample and the required volume of "electrolyte" (5M NaCl) were placed in the beaker, the vessel sealed and purged of air using an inert gas. The vessel was then pressurised with chlorine and allowed to equilibrate (Figure 2a) for approximately 20 hours. It was then decompressed to allow escape of the chlorine in the head space above the solution. This caused the solution to become supersaturated (Figure 2b). The vessel was then resealed and the excess gas in solution released into the head space by bubble evolution (Figure 2c). This resultant pressure rise in the vessel was monitored by a transducer; the slope of the P vs t curve is proportional to the bubble evolution rate, J .[24]

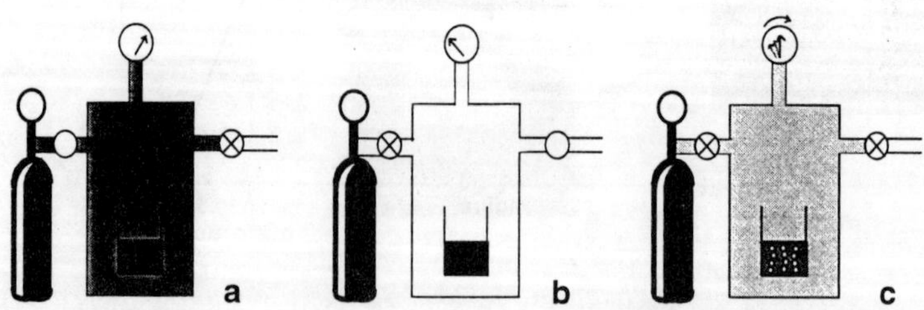

Figure 2. *Schematic diagram of PRBN apparatus; a: during pressurization; b: immediately following decompression; c: some time after decompression. Depth of shading represents chlorine concentration.*

3.2 Electrochemical Quartz Crystal Microbalance (EQCM) Instrumentation

The EQCM technique has been reviewed elsewhere.[25] Application of a voltage across a piezoelectric material, here quartz, induces motion of the crystal. Incorporation of a suitably cut crystal in an oscillator circuit, causes the unloaded crystal to resonate at a frequency f_0. When additional mass is coupled to the resonator, the frequency shifts down from f_0. We used AT-cut quartz crystals, which oscillate in the thickness shear mode.

When one face of the crystal is exposed to a liquid, crystal oscillation launches a shear wave into the liquid. For a Newtonian fluid, the decay length (δ) of the acoustic wave within the liquid is $(\nu/\pi f_0)^{1/2}$, where ν is the kinematic viscosity of the liquid. For the 10 MHz quartz crystals used here in aqueous electrolytes, $\delta \approx 200$ nm. This results in coupling to the electrode of an effective liquid mass per unit area of electrode (ΔM_L) equal to $\delta\rho$, where ρ is the density of the liquid. This in turn results in a shift in resonant frequency:[26]

$$\Delta f = -\left(\frac{2}{\rho_q v_q}\right) f_0^2 \, \Delta M_L \tag{5}$$

where ρ_q is the density of quartz and v_q is the wave velocity within the quartz crystal. In our experiments, we exploit the fact that formation of a bubble at the interface involves displacement of liquid by gas, thereby altering ΔM_L. This is detectable through the associated resonant frequency change (Δf), as illustrated schematically in Figure 3.

EQCM experiments employed a three-electrode, one compartment glass cell. The working electrode, which was also one of the exciting electrodes on the quartz crystal, was Pt. The counter electrode was a Pt gauze. A KCl saturated calomel electrode (SCE) was the reference electrode. The electrolyte was 0.9 M NaCl, containing 0.1 M HCl to minimize local pH variations accompanying hydrogen evolution. Current and frequency measurements were recorded during potential cycling in the range $-0.8 \leq E / V \leq 1.6$ (scan rate: 50 mV s^{-1}).

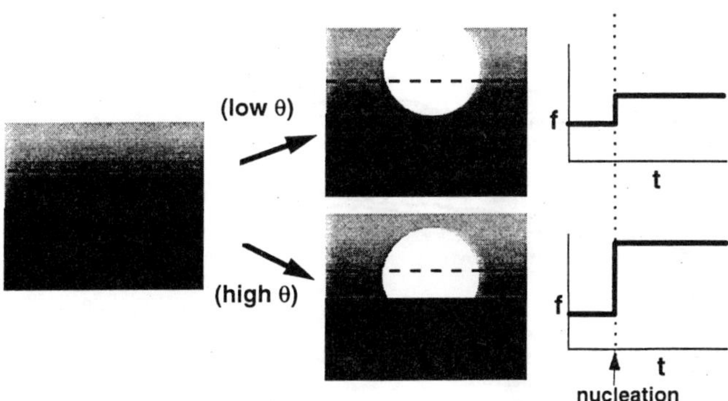

Figure 3. *The EQCM response to bubble evolution. Black represents the electrode, grey the solution, and white a bubble. The horizontal dashed line is a distance δ from the electrode. Upper (lower) frames are for low (high) contact angles, resulting in small (large) liquid displacements and frequency changes. Homogeneous nucleation, beyond δ, results in no frequency change.*

4 RESULTS AND DISCUSSION

4.1 PRBN Studies

Figure 4 shows raw pressure transients following the decompression of chlorine-supersaturated brine solutions from 5.8 bar to 1.5 bar; this corresponds to $\sigma = 2.9$ at $t = 0$, decreasing thereafter as chlorine moves from the liquid to the gas phase. The five traces represent varying RTO formulations.

At a purely operational level, there is clearly an optimum anode formulation for bubble nucleation. We cannot rule out *secondary* physical (morphological) differences between these surfaces, as a consequence of changes in the RTO formulations. Nevertheless, it is reasonable to assume that the differences in surface characteristics are *primarily* chemical. It is therefore interesting that the optimum composition, viewed from the perspective of bubble nucleation, is very similar to that chosen on purely electrocatalytic grounds and used commercially. We therefore have the fortunate situation that it is not necessary to compromise between the generation of dissolved chlorine and its separation via bubble nucleation.

Quantitative data interpretation requires correction for the effect of the adiabatic conditions under which decompression is carried out. Essentially, the rapid decompression at $t = 0$ cools the gas. There are then two contributions to the measured pressure rises of Figure 4. First, there is the transfer of chlorine from the solution to the gas phase as a consequence of bubble nucleation. Second, there is re-heating of the adiabatically cooled gas back to the initial temperature. In order to obtain the former contribution, we need to correct off the latter from the raw data.

This procedure is illustrated in Figure 5. The adiabatic contribution was determined by repeating the experiment in the absence of the electrolyte, so the system was unable to evolve bubbles. The result is curve b in Figure 5. It is the more rapid component of the obviously biphasic raw pressure transient (curve a in Figure 5), and is complete by $t \approx 20$ s. Subtraction of curve b from curve a yields curve c, which we attribute to bubble nucleation.

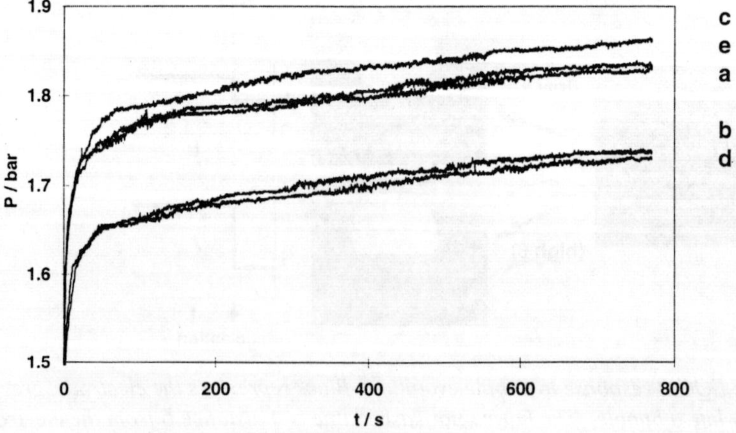

Figure 4. *PRBN pressure transients measured at σ =2.9 on five RTO samples; Ru:Ti ratio increases from a to e.*

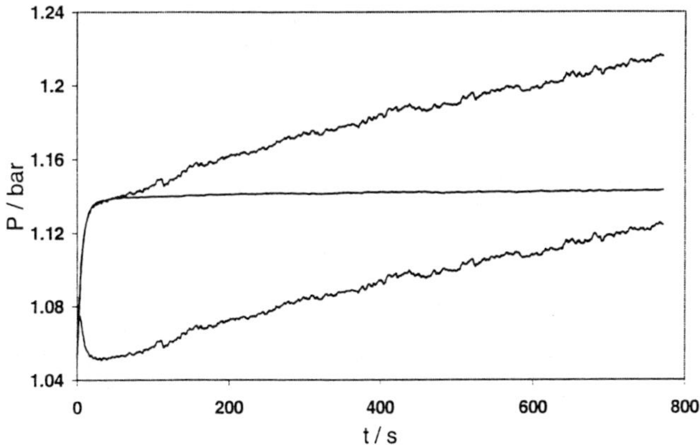

Figure 5. *The correction procedure for the adiabatic cooling and re-heating effect; a: raw data; b: component due to re-heating; c: corrected transient.*

Aside from a minor aberration in the first few seconds, we see a smooth nucleation-driven pressure rise across extended time scales ($t > 1000$ s). The slope of the line (P'), which represents the nucleation rate, decreases steadily with t, as σ decreases progressively during the experiment. Although, in principle, we could determine J from P' as a function of σ by determining instantaneous slopes as a function of time throughout the experiment, the data within a given transient only span a small range of σ. We therefore elected to extrapolate the slopes to short times ($t \approx 20$ s), at which point σ is accurately known from the selected initial and target pressures.

The results of doing this for two series of experiments (corresponding to two temperatures) as a function of supersaturation are shown in Figure 6. Prompted by equation (4), we present the data in the form of ln(J) vs $(\sigma P)^{-2}$ plots, in the anticipation that they will be linear. This is not the case. Rather, we see two limiting slopes at the extremes of $(\sigma P)^{-2}$, with a smooth gradation at intermediate values.

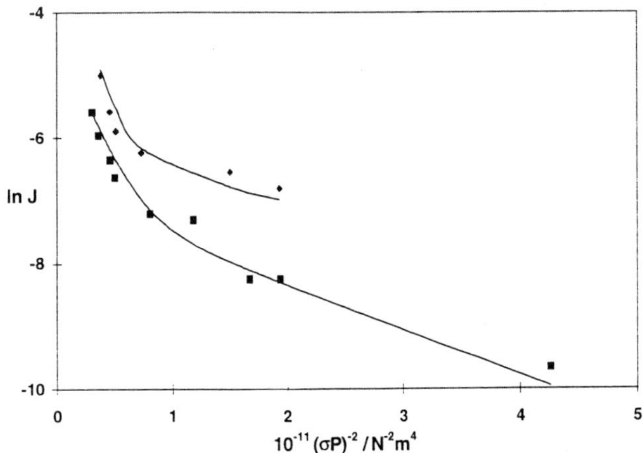

Figure 6. *Plots of equation (4) for chlorine evolution at 25^{0}C (■), 50^{0}C (◆).*

To rationalize this result, we propose that the surface is kinetically inhomogeneous.[27] In this model, we suggest that each kinetically distinct type of site obeys an equation of the form of equation (4), but with different physicochemical parameters. The observed rate is then the sum of the rates at all sites, appropriately weighted for their populations. At low supersaturation (high $(\sigma P)^{-2}$) only the most energetic sites will be active. This corresponds to the region at the right hand side of Figure 6. As the supersaturation is increased (moving to the left in Figure 6) the driving force becomes sufficient to bring progressively less energetic sites into play. The limiting slopes in Figure 6 thus represent the extremes of least and most active sites present.

In the context of the PRBN technique, increasing σ was accomplished by increasing the equilibrating pressure at $t < 0$. In the electrochemical context, this is accomplished by increasing current density. Thus we now turn to such measurements, using the EQCM for bubble detection.

4.2 EQCM Data

The capability of the EQCM to detect bubble evolution is illustrated through Figure 7. In this proof-of-concept experiment we employ a Pt electrode to evolve chlorine and hydrogen, respectively, at sufficiently anodic and cathodic potentials. We are able to make several deductions from the data.

First, and most important, is the obvious point that we are able to detect bubbles at all. This is evident through the frequency increases at the extremes of anodic and cathodic potential. These frequency increases represent decreases in the effective mass coupled to the quartz resonator as gas (chlorine or hydrogen) replaces liquid at the electrode/solution interface. This result unequivocally signals heterogeneous nucleation, vindicating our theoretical approach and consistent with the fact that changing "anode" characteristics in the PRBN experiments does indeed influence bubble evolution rate.

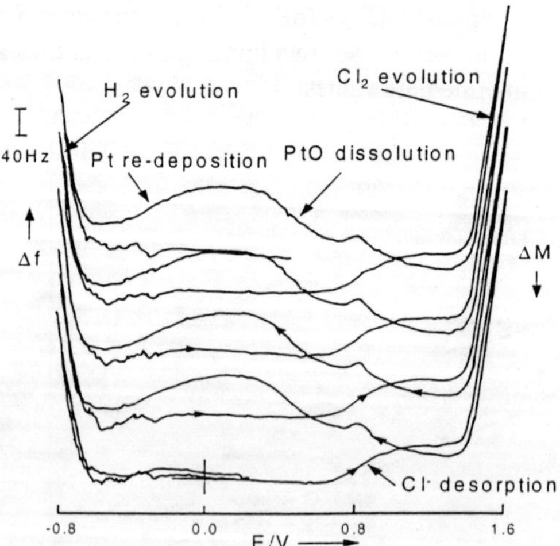

Figure 7. *EQCM frequency response to hydrogen and chlorine bubble evolution. See text for electrode and solution composition. [Reprinted from J. Electroanal. Chem.,* ***335****, F.-B. Li et al., "EQCM studies of potentiodynamic electrolysis of aqueous chloride solution: surface processes and evolution of H_2 and Cl_2 gas bubbles", pp. 345-362 (1992) with kind permission from Elsevier Science S.A., P.O. Box 564, 1001, Lausanne, Switzerland.]*

Second, comparison of the frequency response of Figure 7 with the accompanying current data (not shown) reveals a delay between the onset of current passage and the formation of bubbles. This implies rate-limiting nucleation.

Third, we see that the frequency trace is apparently more noisy in the bubble evolution regime, most obviously during hydrogen evolution. More detailed examination at higher resolution shows that the frequency fluctuations are rather structured. They show a sharp increase, followed by a more gradual increase, then a sharp decrease. We attribute these to nucleation, growth and detachment, respectively, of bubbles. A detailed exploration of these effects for chlorine evolution will be the subject of future study.

Finally, we also see frequency changes in the potential region $-0.4 \leq E / V \leq 1.2$, when bubbles are not being evolved. We do not dwell on this here, since a more detailed discussion has been presented previously.[28,29] However, we note that these changes provide opportunities to study surface processes such as chlorine adsorption, and electrode dissolution and re-deposition, as indicated in Figure 7. The latter processes are the cause of the steady procession of the frequency response up the page (to higher frequency / lower mass). The extreme sensitivity of the EQCM could allow estimation of dissolution-driven limitations to electrode lifetime.

5 CONCLUSIONS

Through pressure-induced supersaturation of brine, the PRBN technique allows qualitative comparison and quantitative characterization of anode coatings as bubble nucleation surfaces. All RTO samples studied showed a distribution of nucleation site energetics, so that the number of sites active in bubble nucleation was a function of supersaturation and pressure. Chlorine bubble nucleation rate is not monotonic with Ru content in the RTO coating: there is an optimum composition, which is fairly similar to that optimized on purely electrocatalytic grounds.

The EQCM provides an alternative means to study bubble evolution, by electrochemical supersaturation of the electrolyte. The ability of the technique to detect bubble evolution unequivocally signals that nucleation is heterogeneous (at the electrode/electrolyte interface) rather than homogeneous (in the bulk solution). In addition to offering the option of gravimetrically monitoring electrode dissolution, the technique promises the opportunity to study nucleation, growth and detachment of individual bubbles.

6 ACKNOWLEDGEMENTS

SLG thanks Leicester University and ICI for a studentship. We thank Dr S.D. Lubetkin for helpful conversations.

7 REFERENCES

1. L.I. Krishtalik, 'Charge Transfer Reactions in Electrochemical and Chemical Processes', Consultants Bureau, New York, 1986, p. 179.
2. A. Shingleton, PhD Thesis, Bath University, 1996.

3. F. Hine, B.V. Tilak and K. Viswanathan, 'Modern Aspects of Electrochemistry', Ed. R. E. White, J. O'M. Bockris and B.E. Conway, Plenum Press, New York, 1996, Vol. 18, p. 249.
4. H. Vogt, *Electrochim. Acta*, 1994, **39**, 1981.
5. B.E. Conway and G. Ping, *Far. Trans.*, 1991, **87**, 2705.
6. S. Trasatti, *Electrochim. Acta*, 1987, **32**, 369.
7. 'Modern Chlor-Alkali Technology', Ed. N. M. Prout and J. S. Moorhouse, Elsevier Science Publishers, Barking, 1990, Vol. 4.
8. G. Kreysa and N. Kuhn, *J. Appl. Electrochem.*, 1989, **19**, 720.
9. D. Pletcher and F. Walsh, 'Industrial Electrochemistry', 2nd ed., Blackie A & P, Glasgow, 1993.
10. B.E. Bongenaar-Schlenter, L.J.J. Janssen, S.J.D. van Stralen and E. Barendrecht, *J. Appl. Electrochem.*, 1985, **15**, 537.
11. R. Cole, *Adv. Heat Transfer*, 1974, **10**, 85.
12. L.E. Scriven, *Chem. Eng. Sci.*, 1959, **10**, 1.
13. S.B. Karri, *Chem. Eng. Comm.*, 1988, **70**, 127.
14. M. Blander and J.L. Katz, *A. I. Ch. E. J.*, 1975, **21**, 833.
15. C.A. Ward, A. Balakrishnan and F.C. Hooper, *J. Basic Eng.*, 1970, **85**, 695.
16. C.A. Ward, P. Yikuisis and A.S. Tucker, *J. Coll. Int. Sci.*, 1986, **113**, 388.
17. P.J. Sides, 'Modern Aspects of Electrochemistry', Ed. R.E. White, J. O'M. Bockris and B.E. Conway, Plenum Press, New York, 1986, Vol. 18, p. 303.
18. 'Controlled Particle, Droplet and Bubble Formation', Ed. D.J. Wedlock, Butterworth-Heinemann, Oxford, 1994.
19. S.D. Lubetkin, *Chem. Soc. Rev.*, 1995, **24**, 243.
20. J.P. Hirth and G.M. Pound, *Prog. Mat. Sci.*, 1963, **11**, 1.
21. P.M. Wilt, *J. Coll. Int. Sci.*, 1985, **112**, 530.
22. S.D. Lubetkin, *Far. Trans.*, 1989, **85**, 1753.
23. D.R. Hodgson, *Electrochim. Acta*, 1996, **41**, 605.
24. M.W. Carr, PhD Thesis, Bristol University, 1993.
25. S. Bruckenstein and A.R. Hillman, 'Handbook of Surface Imaging and Visualization', Ed. A.T. Hubbard, CRC Press, Boca Raton, 1995, p.101.
26. G.Z. Sauerbrey, *Z. Phys.*, 1959, **155**, 206.
27. M.W. Carr, A.R. Hillman and S.D. Lubetkin, *J. Coll. Int. Sci.*, 1995, **169**, 135.
28. M.W. Carr, A.R. Hillman, S.D. Lubetkin and M.J. Swann, *J. Electroanal. Chem.*, 1989, **267**, 313.
29. F.-B. Li, A.R. Hillman, S.D. Lubetkin and D.J. Roberts, *J. Electroanal. Chem.*, 1992, **335**, 345.

9

CO-DEPOSITED COMPOSITE LAYERS OF Ni+RuO_2 FOR ELECTROCATALYSIS OF HYDROGEN EVOLUTION

Ana B. Tavares* and S. Trasatti

Department of Physical Chemistry and Electrochemistry
University of Milan, Milan, Italy

1 INTRODUCTION

Ni is a most appropriate material for technological applications in strongly alkaline media. As a cathode, however, under intense hydrogen evolution, Ni activity decreases with time. Recently,[1] oxides have been shown to possess interesting electrocatalytic activity for hydrogen evolution with only moderate decays in performance with use. Among these, RuO_2 is one of the most active. Therefore, RuO_2-activated Ni electrodes could offer a decisive improvement in performances. Although small amounts of RuO_2 are required to activate Ni, the deposition of the oxide by thermal decomposition does not produce mechanically stable electrodes due to erosion by evolving gas bubbles.

Various attempts have been made to obtain mechanically more stable activated electrodes. Among these, particularly attractive is the co-deposition of the active component embedded in a solid matrix.[2] The aim is to obtain a layer containing a very small amount of the electrocatalyst firmly held by the co-deposited solid phase so that the active particles are not swept away by the evolving gas bubbles. Both cathodes[3-5] and anodes[6] can be prepared, although the latter are for the time being much less investigated.

Composite layers of Ni+RuO_2 co-deposited on a Ni support have been prepared for the first time by Iwakura *et al.*[7] They have verified that the activated cathode exhibits improved performance with respect to pure Ni. However, these authors have not characterised their electrodes in many details.[8] Further, they have used only one kind of electrodeposition bath.

In this work, RuO_2 powders prepared by thermal decomposition of $RuCl_3$ have been suspended in a Ni galvanic bath and co-deposited onto a Ni support. Various parameters have been investigated: (1) the pre-treatment of the support (polishing *vs.* etching), (2) the composition of the bath (sulphate *vs.* chloride *vs.* thiosulphate), and (3) the amount of RuO_2 in suspension.

2 EXPERIMENTAL

RuO_2 powders were obtained by thermal decomposition of $RuCl_3{\cdot}nH_2O$ at 350 °C in air.

*Permanent address: Dep. Quimica, Fac. Ciencias, Univ. Lisboa, 1700 Lisboa (Portugal).

After 1 hour the powder was crushed and milled in an agata mortar and kept in the oven for 1 additional hour. The operation was repeated three times. The formation of RuO_2 was confirmed by X-ray diffraction (XRD) analysis.

Ni platelets of 10×10×0.2 mm size (with a very thin stem) were used as a support. The support was pretreated in two ways giving opposite surface morphologies: (a) mechanical polishing by means of emery-paper down to 800 mesh, producing very smooth surfaces, and (b) chemical etching in boiling aqueous HCl resulting in very rough surfaces. Although rough surfaces may offer better adherence to the electrodeposited layer, electrodeposition conditions may be more reproducible on a smooth surface.

The RuO_2 powder was dispersed in the electrodeposition bath (100 ml) and kept in suspension by mechanical stirring. The Ni electrode was immersed in the centre of a Raney-Ni cylindrical mesh working as a counter-electrode. No attempt was made at this stage of the work to optimise the geometry of the cell for a better control of the homogeneity of the electrodeposition reaction.

The composition of the bath was varied to investigate the effect of the quality of the Ni deposit. Ni galvanic baths can be of various types depending on the purpose of the operation. The most classical bath is the so-called Watt's bath whose main component is $NiSO_4$. However, in the literature composite Ni-based electrodes have been prepared also from Ni baths whose prevailing component is $NiCl_2$.[9] Further, co-deposition of Ni and suspended transition metal sulphides has been carried out from baths containing some thiosulphate.[10,11] Sulphur-containing species are known to be active for hydrogen evolution.[12]

In a first experiment, the amount of RuO_2 in suspension was varied up to 50 g dm^{-3}. The activity of the electrode increased with the RuO_2 content reaching a saturation value already for 10 g dm^{-3}. Thus, most of the variation in the activity takes place up to 10 g dm^{-3} of RuO_2. For this reason, in successive experiments attention was focused onto this bath composition range. It appears in fact that much less than 10 g dm^{-3} RuO_2 is needed to achieve maximum electrode activation.

The electrodeposition was carried out at 50 °C with a deposition time of 1 hour at a current density of 5 mA cm^{-2}. Although these parameters were not optimised in great detail, these values represent conditions close to the best efficiency; an increase in current density, in particular, produced a decrease in Ni deposition efficiency.

3 RESULTS AND DISCUSSION

3.1 Efficiency of Electrodeposition

A first clear difference between the various Ni baths has been observed in the variation of the amount of electrodeposited Ni at constant amount of charge with the amount of RuO_2 in solution. In the case of sulphate and chloride baths the current yield of Ni decreases with increasing amount of RuO_2. In the case of the thiosulphate bath, the current yield increases. A possible explanation is that in the former case the deposition of RuO_2 decreases the overpotential for hydrogen evolution[1] which in turn depresses the Ni electrodeposition reaction. In the latter case, it is found that sulphur is co-deposited with Ni: the presence of RuO_2 reduces the current lost in the reduction of sulphur species.

3.2 Surface Morphology

The morphology of the electrodeposited layers of pure Ni was observed on the SEM. While sulphate and chloride baths produce crystalline Ni deposits, the morphology of the Ni layers from the thiosulphate baths is totally different, being more similar to what is obtained by an electroless Ni deposition. The layer consists of nodules like a floor of cobble-stones.[11]

That Ni sulphide is formed during Ni deposition in the presence of thiosulphates is clearly shown by the chemical analysis by EDX.

3.3 Deposition of RuO_2

The presence of RuO_2 on the surface of Ni is revealed on the SEM by small white particles. In all cases an increase in roughness is apparent, which is highest for the sulphate bath.

The deposition of RuO_2 has been confirmed by several approaches: X-ray analysis, EDX analysis, and chemical analysis, the last consisting of dissolving the support in boiling HCl and weighing the residue. The amount of RuO_2 deposited with Ni has thus been found to be up to 0.5 mg cm^{-2}. The distribution of Ru on the surface has been mapped by EDX: it has turned out to be homogeneous, although the unit of RuO_2 species is as a rule a particle or aggregate of particles. SEM has revealed that these particles adhere to the Ni matrix: however, a profile of RuO_2 across the layer could not be determined.

A strict quantitative analysis of the RuO_2 content in the layer as a function of the RuO_2 content in the bath could not be accomplished. However, an attempt was made on the basis of the X-ray spectra: the results showed that the amount of RuO_2 increases and reaches a value close to saturation already at about 2 g dm^{-3} of oxide in suspension.

3.4 Voltammetric Curves

The electrochemical surface properties of these composite electrodes were investigated by means of cyclic voltammetric curves in 1 mol dm^{-3} KOH solution. In all cases the curves showed the characteristic peaks of Ni. In the case of the thiosulphate bath the voltammetric curve exhibited some extra features attributable to the presence of sulphur-containing species (Figure 1).

The surface amount of RuO_2 is not sufficiently high as to produce detectable features in the shape of the voltammetric curves. Nevertheless, the presence of RuO_2 is revealed by the increasing ratio between the current in the so-called double-layer region and the current at the Ni peaks (Figure 2). A semi-quantitative relative estimation of the surface content of RuO_2 can thus be attempted.[13]

3.5 Voltammetric Charge

Integration of the voltammetric curves gives the voltammetric charge which is proportional to the surface amount of active sites exchanging charge with the solution during a voltammetric cycle.[14] The voltammetric charge has been found to increase with the amount of RuO_2 in solution reaching a saturation value already at about 2 g dm^{-3} (Figure 3). While the sulphate and the chloride baths gave comparable voltammetric charges, the thiosulphate bath produced higher charges. This is presumably due to redox transitions of the sulphur species and in addition to the nanocrystallinity of the deposit.

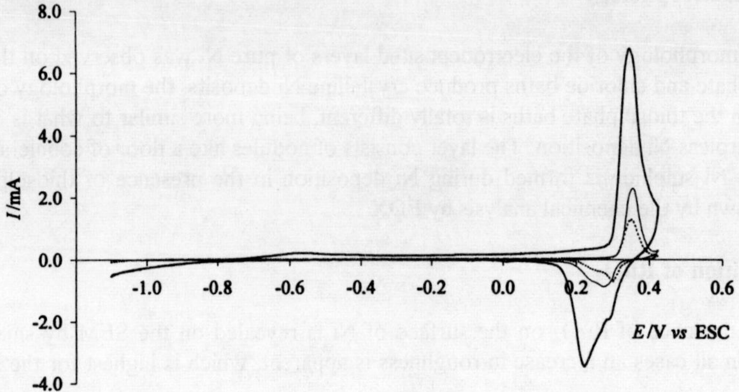

Figure 1 *Voltammetric curves at 20 mV s⁻¹ of electrodeposited Ni layers from three different baths. (·····) Chloride; (———) Watts; (———) Thiosulphate bath.*

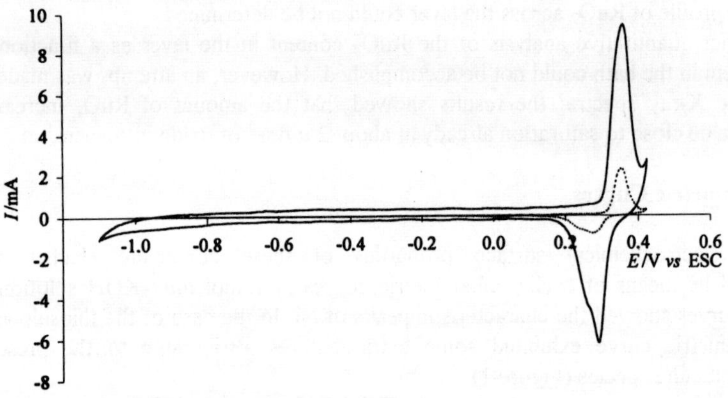

Figure 2 *Voltammetric curves at 20 mV s^{-1} of an electrodeposited (·····) Ni layer and (———) Ni + RuO_2 layer (10 g dm^{-3}). Watts bath; etched support.*

3.6 Polarization Curves

The electrocatalytic activity for hydrogen evolution has been investigated by recording quasi-stationary current-potential curves and by determining the reaction order with respect to OH^-. The polarization curves have been found to be characterized by a single Tafel line in the explored current density range (up to *ca.* 50 mA cm^{-2}) (Figure 4). Different behaviour has been observed with the electrodes prepared from the thiosulphate bath. In the case of the other two baths an appreciable hysteresis is found between the forward and the reverse potential scan which becomes more evident as the amount of RuO_2 increases. This is a typical feature of RuO_2 which dominates the surface properties.[1] With the thiosulphate bath, the reverse scan shows a dramatic modification of the kinetic

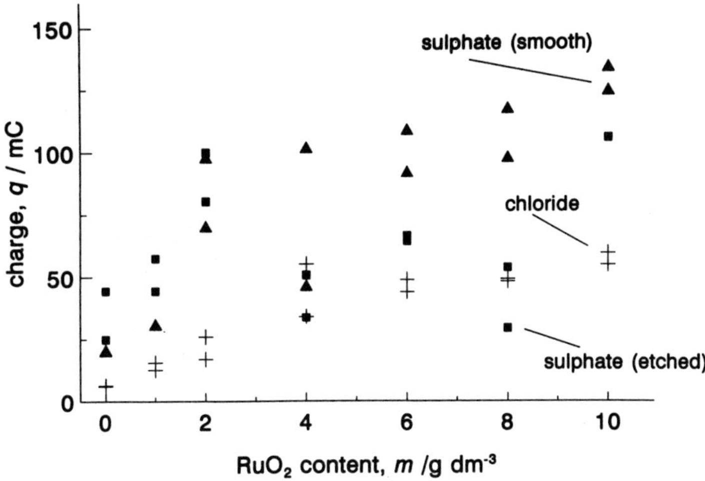

Figure 3 *Dependence of the voltammetric charge on the amount of RuO_2 in the baths. (▲) Watts bath (sulphate), smooth support; (■) Watts bath (sulphate), etched support; (+) Chloride bath, etched support.*

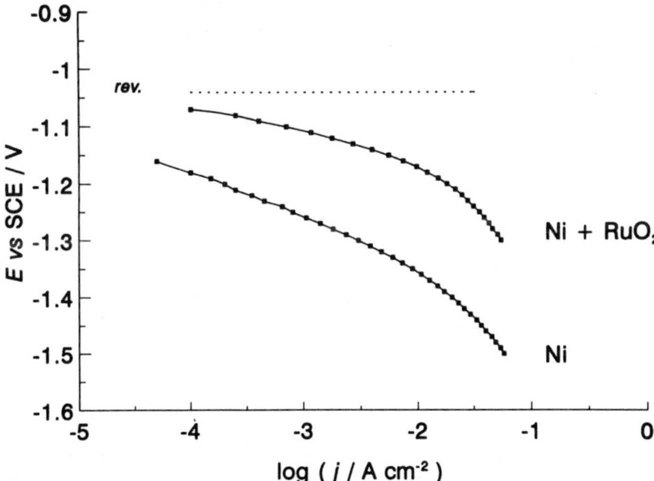

Figure 4 *Typical Tafel lines (backward scan) for Ni and Ni + RuO_2 electrodes (10 g dm^{-3}). Watts bath (sulphate), smooth support. (·····) Hydrogen reversible potential.*

parameters presumably related to hydrogenation of surface sulphur species.

The Tafel slope is *ca.* 100 mV for pure Ni from sulphate and chloride baths, a typical value for this metal[12] (Figure 5). As soon as a minimum amount of RuO_2 is added, the Tafel slope begins to decrease, settling off around 60 mV at higher RuO_2 contents. However,

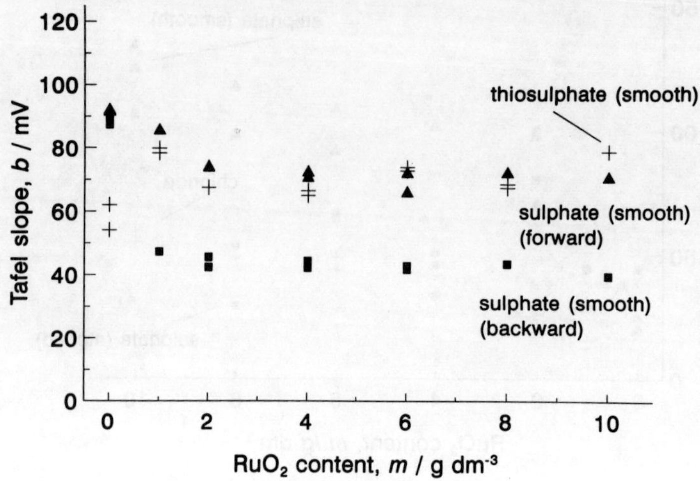

Figure 5 *Tafel slope for the hydrogen evolution reaction as a function of the RuO_2 content in the bath. Smooth support. (▪) Watts bath (sulphate), backward scan; (+) Thiosulphate bath; forward scan; (▲) Watts bath (sulphate), forward scan.*

there is a marked difference between polished and etched electrodes: in the latter case the decrease of the Tafel slope is little apparent despite the indisputable presence of RuO_2 on the surface. A possible explanation is that with etched electrodes Ni dominates even in the presence of RuO_2 because of the much higher extension of its surface area.

In the reverse scan the Tafel slope is lower, close to 40 mV. This activation is typical of RuO_2 electrodes[1,15] and shows that the features of this material come out better on a surface under intense hydrogen evolution. During the reverse potential scan any difference due to the pretreatment of the support tends to vanish. Quite surprisingly, the picture is opposite with the thiosulphate bath. The pure Ni electrodes show in fact a lower Tafel slope than the RuO_2 activated ones. This explains the observed behaviour of the Ni yield (see above). The Tafel slope becomes lower during the reverse scan also in this case, but a precise determination of its value was not possible.

3.7 Electrocatalytic Activity

A plot of current at a given potential as a function of RuO_2 in solution (Figure 6) reveals that *j* varies as the voltammetric charge does. Since the voltammetric charge parallels the surface concentration of RuO_2, this is also the case of activity (Figure 7). Thus, the mechanism of the electrocatalysis does not appear to change with the surface amount of RuO_2. A change is observed only as RuO_2 is added to the solution, *i.e.* RuO_2 is introduced into the Ni deposit.

3.8 Stability

Voltammetric curves were recorded again after the kinetic study of hydrogen

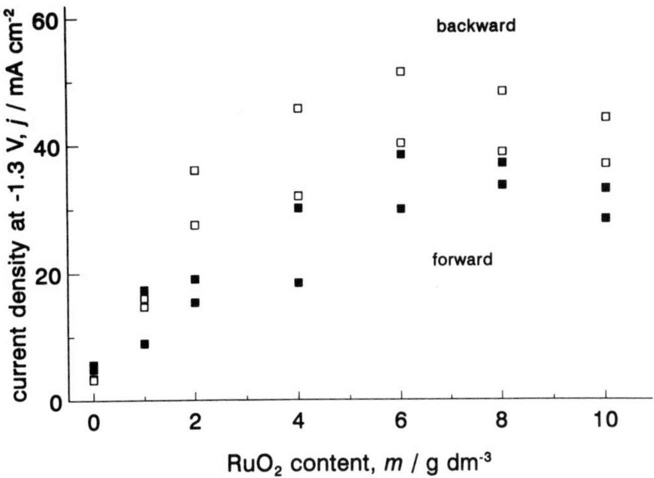

Figure 6 *Current density at -1.3 V vs SCE as a function of the amount of RuO$_2$ in the bath. Chloride bath. (▪) Forward scan; (□) Backward scan.*

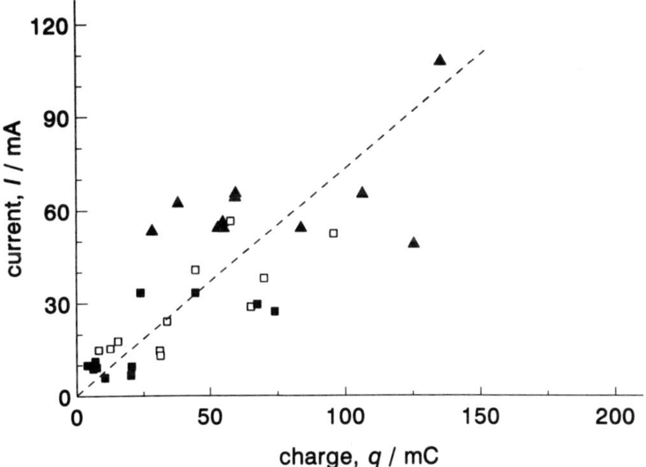

Figure 7 *Current of hydrogen evolution at -1.3 V vs SCE as a function of the voltammetric charge of electrodes obtained from different suspension composition. (▪) 0; (□) 1; (▲) 10 g dm^{-3} RuO$_2$.*

evolution and were compared with the initial values. For sulphate and chloride baths the voltammetric charge remains unaltered (Fig. 8) which indicates that the surface properties are not changed by hydrogen evolution. In the case of the thiosulphate bath, the charge after the hydrogen evolution study turns out to be higher by one order of magnitude (Fig. 9).

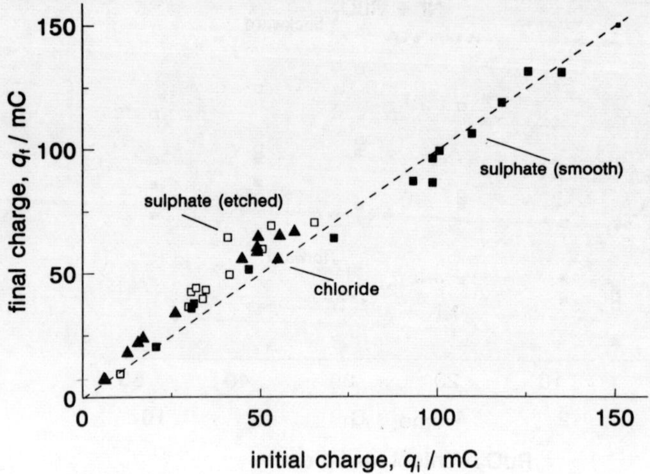

Figure 8 *Voltammetric charge after the hydrogen evolution study plotted against the charge for fresh electrodes. (▲) Chloride bath; (■) Watts bath (sulphate).*

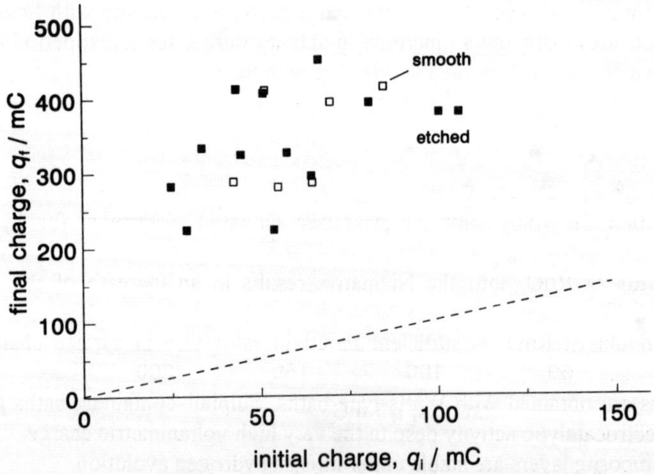

Figure 9 *Same as in Figure 8. Thiosulphate bath. (■) Etched support; (□) Smooth support.*

This can be understood in terms of surface modifications resulting in a very subdivided Ni surface. It is intriguing that an increase in activity does not correspond to the dramatic increase in charge observed in the case of the thiosulphate bath. This indicates that the surface charge in the latter case cannot be compared with that of the other two baths presumably because of different kinds of surface reactions. This outcome can help

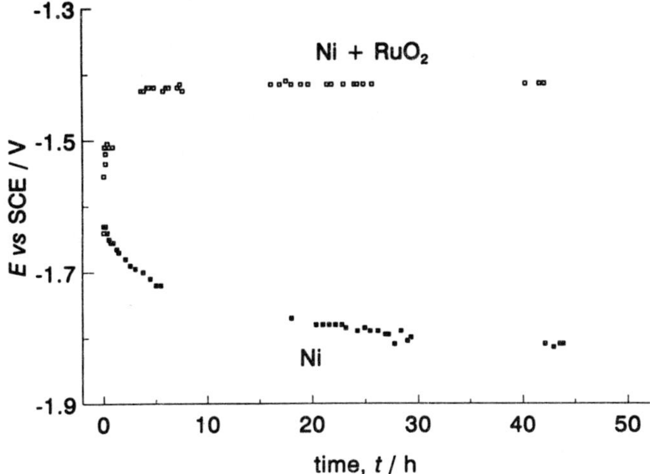

Figure 10 *Electrode potential at 100 mA cm^{-2} as a function of polarization time. Watts bath; smooth support. (▪) Ni; (□) Ni + RuO$_2$ (10 g dm^{-3}).*

understand the increase in Ni electrodeposition efficiency with increasing amount of RuO_2 (*cf.* sec. 3.1). Thus, the hydrogen evolution reaction turns out deactivated rather than activated.

The electrodes were subjected to hydrogen evolution at 100 mA cm^{-2} for 48 h to check their long-term performance. Pure Ni showed a decrease in activity with time, while RuO_2-activated electrodes exhibited an increase in activity during the initial period (Figure 10). No decrease in activity was observed in this case with further use.

4 CONCLUSIONS

(a) The co-deposition of RuO_2 with Ni produces activated electrodes for hydrogen evolution.
(b) The incorporation of RuO_2 into the Ni matrix results in an increase of the surface roughness.
(c) A minimum amount of RuO_2 is sufficient to attain saturation in surface charge and surface activity.
(d) The best results are obtained with Watts-type baths. Sulphur-containing baths give on the whole lower electrocatalytic activity despite the very high voltammetric charge.
(e) (RuO_2 + Ni) composite layers are stable under intense hydrogen evolution.

Acknowledgements.

A.B. Tavares thanks JNICT Portugal (Program PRAXIS XXI) for the Ph.D. grant (BD/2648/93-IF). S. Trasatti is grateful to C.N.R. (Rome) for financial support.

References

1. S. Trasatti, 'Modern Chlor-Alkali Technology', T. C. Wellington Ed., Elsevier, Amsterdam, 1992, Vol. 5, p. 281.
2. A. Hovestad and L. J. J. Janssen, *J. Appl. Electrochem.*, 1995, **25**, 519.
3. H. Dumont, P. Los, A. Lasia, H. Ménard and L. Brossard, *J. Appl. Electrochem.*, 1993, **23**, 684.
4. O. Savadogo, *Electrochim. Acta*, 1992, **37**, 1457.
5. A. Anani, Z. Mao, S. Srinivasan and A. J. Appleby, *J. Appl. Electrochem.*, 1991, **21**, 683.
6. M. Musiani, *J. Chem. Soc. Chem. Commun.*, 1996, 2403.
7. C. Iwakura, N. Furukawa and M. Tanaka, *Electrochim. Acta*, 1992, **37**, 757.
8. C. Iwakura, M. Tanaka, S. Nakamatsu, H. Inoue, M. Matsuoka and N. Furukawa, *Electrochim. Acta*, 1995, **40**, 977.
9. O. Savadogo and C. Allard, *J. Appl. Electrochem.*, 1991, **21**, 73.
10. A. Nidola and R. Schira, *Int. J. Hydrogen Energy*, 1986, **11**, 449.
11. L. B. Albertini, A. C. D. Angelo and E. R. Gonzalez, *J. Appl. Electrochem.*, 1992, **22**, 888.
12. S. Trasatti, 'Advances in Electrochemical Science and Engineering', H. Gerischer and C. W. Tobias Eds., Vol. 2, VCH, Weinheim, 1992, p. 1.
13. C. Angelinetta, S. Trasatti, Lj. D. Atanasoska and R. T. Atanasoski, *J. Electroanal. Chem.*, 1986, **214**, 535.
14. S. Trasatti, 'Electrochemistry of Novel Materials', J. Lipkowski and P. N. Ross Eds., VCH, Weinheim, 1994, p. 204.
15. I. M. Kodintsev and S. Trasatti, *Electrochim. Acta*, 1994, **39**, 1803.

10

THE INFLUENCE OF DIAPHRAGM STRUCTURE AND ACTIVATED CATHODES ON THE PERFORMANCE OF THE CHLORINE ELECTROLYSER

Ginter Nawrat, Maciej Gonet, Teresa Buczek-Karczewska, Andrzej Małachowski, and Adam Korczyński

Institute of Chemistry, Inorganic Technology and Electrochemistry,
Silesian Technical University,
ul. Bolesława Krzywoustego 6, 44-100 Gliwice,
Poland

1 ABSTRACT

This paper presents the results of the investigations and tests done with the cathode-diaphragm system composed of the modified asbestos diaphragm of anisotropic structure and the cathode with an active coating. The coatings were obtained using three methods: by electroplating, chemical and thermal. Amongst the active coatings that have been investigated those based on nickel and cobalt with the addition of platinum, obtained by thermal decomposition of corresponding metal salts on the nickel substrate showed the lowest hydrogen overpotential. Anisotropic diaphragms feature higher porosity in the cathode-adjacent layer as compared to the remaining part of total diaphragm volume and therefore a lower susceptibility to the precipitation within this layer of sparingly soluble calcium and magnesium compounds originating from brine. This results in increased electric conductivity, permeability to brine, and most of all contributes to longer operational life of diaphragms. The use of the proposed cathode-diaphragm system in the experimental cell under test condition of the diaphragm-cell process for the electrolysis of sodium chloride enables cell terminal voltage to be reduced by ca 0.3 V which corresponds to a reduction in specific power consumption of 8 – 10 % and ensures the lengthening of the diaphragm operational life by ca 30 % as compared with conventional diaphragms.

2 INTRODUCTION

The operational life of activated titanium electrodes ranges between 7 and 10 years and their low chlorine overpotential renders any further improvement of the diaphragm electrolysis process technical-economic performance parameters almost only possible by improving the quality of diaphragms and cathode material.

A relatively rapid decline in the permeability to brine and increase in ohmic resistance of diaphragms lead to a worsening of performance parameters of the electrolysis process and consequently make diaphragm replacement necessary after about one year of operation.[1] This result is caused by impurities settling in diaphragms during industrial operation. Firstly, brine impurities, in the form of sparingly soluble calcium and magnesium compounds, precipitate in the cathode-adjacent, alkaline zone of the diaphragm.[2] Diaphragm pores are then gradually filled with calcium and magnesium compounds, with the precipitate accumulating in the direction opposite to the solution flow, i.e. progressing from the catholyte- to the anolyte-adjacent side, which leads to a change in hydrodynamic conditions within the diaphragm. This is accompanied by the brine hydrostatic head increase and the local differences in diaphragm structure and thickness become more and more distinct. The fraction of diaphragm areas of the lowest thickness participating in the overall brine flow at the flow rate required by the process increases which leads to diaphragm alkalisation, first at the thickest spots, and then within the entire diaphragm volume.[3] In order to avoid the deterioration of the cell performance parameters caused by these phenomena a concept of anisotropic diaphragms was conceived. Such diaphragms are characterised by zonal differences in active porosity with the latter increasing in the direction of solution flow. This should lessen the unfavourable tendency of sparingly soluble compounds to settle inside the pores, because in such diaphragms the pore size rises with the increase in the size of particles settling along the path of brine flow.[4] So the use of anisotropic diaphragms in the brine electrolysis process should result in longer repair-to-repair cell operation periods due to the prolonged diaphragm life, as well as in the improvement of the electrolysis process performance parameters, and in addition, lower asbestos consumption which is very important from an ecological point of view.

Further improvement of the process performance parameters, resulting from the reduction in power consumption by 10 – 15%,[5] can be achieved by improving electrocatalytic properties of the cathode material. This improvement is possible by depositing an active coating exhibiting low hydrogen overpotential, high corrosion resistance and electrochemical properties stable under the conditions of the cathodic process of sodium chloride electrolysis. Particularly interesting, in this respect, are alloys of metals with different types of electron configuration, with one of them strongly adsorbing hydrogen,[6] as well as amorphous alloys of metal-metal or metal-metalloid type, the latter being usually boron or phosphorus.[7] At the same time, the use of nickel and cobalt based coatings deposited - in order to ensure proper corrosion resistance of the electrode - on suitably prepared nickel substrate is known to be advantageous in this instance.[8]

3 EXPERIMENTAL

Modified asbestos diaphragms of anisotropic structure and nickel electrodes with nickel and cobalt based active coatings, with both the diaphragm and the electrode constituting an integrated cathode-diaphragm system, were the subject of the present study and underwent tests in the experimental electrolysis cell on a laboratory scale.

3.1 Anisotropic Diaphragms

In the study asbestos diaphragms were investigated in which a desired anisotropy of the structure was obtained in the following arrangements:

– double-layer with the cathode-adjacent layer made of long-fibre and the anode-adjacent one of short-fibre asbestos with the addition of a modifier in both the layers,
– single-layer made of long-fibre asbestos with the addition of a modifier and with the anode side surface sealed with fine-fibre asbestos.

Diaphragm samples were prepared under conditions strictly approximating those of the modified asbestos diaphragm industrial fabrication process and consisted of:

– preparing asbestos slurry with an addition of a modifier,
– diaphragm depositing,
– drying and thermal forming (baking).

In the process of diaphragm deposition a slurry of asbestos in catholyte containing 14 g fibre/dm^3 was used. Three chrysotile asbestos grades were used in the study: long-fibre P-3-50, short-fibre P-4-20 and fine-fibre P-6-40 – all of them produced in Russia.

The following modifiers, made of polytetrafluorethylene (PTFE), were used: fibrous SM-1 grade and fibrous-granular SM-2 grade from OxyTech Systems, Inc. and granular MTM grade from Zakłady Azotowe Tarnów, Poland, added in amount of 20 % wt in relation to the entire diaphragm weight. Another modifier used was HALAR – a copolymer of ethylene and chlortrifluorethylene (ECTFE) from Ausimont USA, Inc. – in amount of 5 % wt. A surface-active agent SM-0 was also used in the diaphragm preparation, also from OxyTech Systems, Inc. Diaphragms

75 x 130 mm in size and of 1.5 kg/m^2 surface density were deposited using the vacuum method with negative pressure changed gradually from 0.02 to 0.07 MPa. The first long-fibre asbestos layer was deposited at a negative pressure of 0.02 – 0.04 MPa. The second one was applied onto the first immediately after the latter had been initially dried by sucking off the solution under 0.04 – 0.07 MPa vacuum. After having been dried with the air being sucked, first at an ambient temperature and then at 95 °C, the diaphragms were subjected to thermal forming at 365 °C – in the case of the diaphragms modified with SM-1, SM-2 or MTM, or at 255 °C – in the case of the diaphragms modified with HALAR. In the thermal forming process the diaphragms were heated from 100°C up to the sintering temperature at 1 °C/min and then held at the sintering temperature for about 60 min.

The diaphragm samples obtained were then subjected to the tests of permeability to brine of 300 – 305 g NaCl/dm^3 concentration, at 80 °C and at water head of 300 mm and then ohmic resistance was measured using 1000 Hz sinusoidal alternating current and a generator-controlled potentiostat as a stable electric current source. The measurements of the diaphragm effective ohmic resistance were performed in the saturated sodium chloride solution at pH=7 and at 80 °C after previous diaphragm soakage (seasoning) for at least 24 hours. From the diaphragm electrical resistance measurement results, the MacMullin number[9] and theoretical current efficiency of the electrolysis process as determined on the basis of the diaphragm cell mathematical model,[10] were calculated.

3.2 Activated Cathodes

The electrodes to be examined were obtained by depositing active coatings on a nickel substrate. The following alloy coatings were tested:

- nickel-molybdenum, nickel-vanadium and nickel-molybdenum-vanadium prepared by electroplating,
- nickel-phosphorus and cobalt-phosphorus obtained by chemical method,
- selected two-component systems, based on nickel and cobalt in combination with platinum, palladium, iridium and ruthenium, obtained by thermal method.

In the investigations, nickel-substrate electrodes were used in the form of rods, 3 mm in diameter and 50 mm in length, and in the form of 64 mesh/cm^2 (400 mesh/inch^2) micronet made from 0.1 mm diameter wire.

The coatings were deposited on degreased and etched substrates by electroplating, chemical or thermal method. Plain and complex baths were used in the electroplating process and the preliminary tests revealed that plain baths produced coatings of unsatisfactory quality. Alloy coatings, showing good adherence to the substrate, without any visible cracks or losses, of more than 10 μm thickness were obtained from complex baths. The Ni-Mo, Ni-V and Ni-Mo-V alloy coatings were obtained from tartrate baths under constant current density conditions.

Each coating to be tested was deposited from a fresh electroplating bath containing:

- nickel(II) sulphate(VI), $NiSO_4 \cdot 7H_2O$ 62.0 g/dm^3,
- potassium-sodium tartrate, $KNaC_4H_4O_6 \cdot 4H_2O$ 76.0 g/dm^3,
- sodium chloride, NaCl 18.0 g/dm^3,
- ammonia aqueous, $NH_3 \cdot aq$ up to pH = 10.0.

Alloying additions were introduced into the bath in the form of their salts, using correspondingly:

- sodium molybdenate(VI), $Na_2MoO_4 \cdot 4H_2O$ 15.3 g/dm^3,
- vanadium(IV) oxide sulphate(VI), $VOSO_4 \cdot H_2O$ 5.0 g/dm^3.

The electroplating process was performed under the following conditions:

- bath temperature 30 °C,
- cathodic current density 0.5 – 1.0 kA/m^2,
- pH range maintained 9.8 – 10.1.

In the process of chemical coating of the electrode surface, it was subjected to a sensitising and activating treatment (sensactivation) in a solution of tin(II) and palladium(II) salts and a metallizing treatment in a solution containing sodium phosphinate (hypophosphite) and nickel(II), or cobalt(II) ions, at pH = 3 – 10.

Acid and alkaline baths containing:

– nickel(II) chloride, $NiCl_2 \cdot 6H_2O$ 25 – 35 g/dm^3

– sodium phosphinate, $NaH_2PO_2 \cdot H_2O$ 16 – 22 g/dm^3,

were used. A succinate buffer solution was used in acid whereas citrate-ammonia was used in alkaline baths. The process was carried out at 75 °C for 1 – 10 hours, with the duration lengthened with the solution acidity increase.

Solutions of inorganic salts of corresponding metals in n-butanol were used in the process of thermal coating deposition. The solutions were applied onto a pre-treated nickel substrate with a paintbrush. Coatings consisting of 3 – 15 layers were obtained in this way. Each layer was heat treated for 15 min in atmospheric air and after all the layers had been deposited, the electrode was further heated for 60 min. The electrodes were thermally treated within a temperature range of 300 – 450 °C. The coatings obtained by the thermal method were then subjected to a cathode reduction process at 80 °C, in 1M NaOH solution, for 1 – 25 h, using a current of 5 kA/m^2 density.

The electrode surface morphology was examined under the scanning electron microscope and the chemical composition was determined with the microscope coupled with the energy dispersion spectrometer. Phase analysis was done using the X-ray diffractometer. Electric capacitance of the double layer was determined from the electrode impedance measurement results. Hydrogen evolution overpotential was evaluated on the basis of the polarisation curves plotted with the potentiostatic set from Tacussel.

Double layer capacitance measurements were performed in 5M NaCl solution, at 25 °C, within a potential range from -0.95 to -0.45 V versus SCE (saturated calomel electrode). Hydrogen evolution overpotential was measured in 1M NaOH solution, at 25 °C, within a potential range from -1.90 to -1.00 V vs SCE.

3.3 Cathode-Diaphragm Systems

The cathode-diaphragm systems obtained were tested in experimental cells on a laboratory scale. The cathodes were previously activated by attaching a nickel micronet having an active coating or by depositing a layer of carbon fibres activated in a similar manner as the nickel micronet.

The tests were performed using a laboratory set of two electrolysis cells made of acrylic resin and fitted with DSA anodes. The cell height was ca 800 mm which enabled the level difference between the anolyte and catholyte to be maintained at 500 mm. The cathode was in the form of a typical steel wire net used in industrial electrolysis cells. Cathodic current density was 1.4 kA/m^2.

The tests were carried out under conditions similar to those of the industrial process, with the exception of calcium and magnesium ion impurity levels, which were ca 100 times higher as compared to the typical process brine. Full electrolysis cycle duration was 8 h. The catholyte and anode gas compositions were checked every hour. The electrolysis process consisted of two stages:

– the first, lasting for ca 2 h, was aimed at reaching stable operating conditions of the diaphragm and therefore pure brine was used containing ca 305 g NaCl/dm^3 at pH=3,

– the second, during which the influence of the initial characteristics of the diaphragm on its operational features in a test brine containing additionally 0.4 g Ca^{2+}/dm^3 and 0.1 Mg^{2+}/dm^3 were to be determined.

Parameters of the test electrolysis process were as follows:

– current 10 A,

– anodic current density 1.8 kA/m^2,

– cathodic current density 1.4 kA/m^2,

– brine temperature 80 °C,

– brine input flowrate 0.15 dm^3/h,

– specific brine input flowrate 15.4 $dm^3/dm^2 \cdot h$.

4 DISCUSSION

4.1 Anisotropic Diaphragms

The basic feature of conventional diaphragms obtained in a single procedure of depositing homogeneous mixture of all necessary components is an almost homogeneous structure over the entire diaphragm cross-section. Also, even a major change in the short- to long-fibre asbestos ratio affects their structure only slightly. Much better from this point of view are the diaphragms obtained in a two-step operation of component deposition with the first long-fibre asbestos layer being deposited directly onto the cathode followed by the second, short-fibre layer. This method gives considerably higher volume of voids (unfilled internal spaces) and less compacted fibre structure in the cathode-adjacent layer which can be seen on the microphotograph (Figure 1). This loose structure becomes more and more dense with the decreasing distance to the anode-adjacent surface where the fibre structure is most compact and at the same time the voids volume is smallest.

Both the physico-chemical properties and the results of the electrolysis test unequivocally indicate that the operational properties of anisotropic diaphragms are much better than those of the conventional ones. Anisotropic diaphragms exhibit better permeability to brine and higher electric conductivity (Figure 2).

The compositions of anode gas and catholyte also are improved which results in higher current efficiency which is particularly distinct in the second phase of the test electrolysis, i.e. when using calcium and magnesium ion-doped brine (Figure 3).

This corroborates lower anisotropic diaphragm susceptibility to detrimental effects of impurities which improves both the technical and economic performance parameters of the electrolysis process and promises longer repair-to-repair times of chlorine cells. This effect is achieved thanks to the looser structure in the cathode zone of anisotropic diaphragms.

The properties of anisotropic diaphragms depend substantially on the modifier used. SM-1 fibrous modifier-based diaphragms show the highest permeability to brine and electric conductivity, contrary to the MTM granular modifier-based diaphragms which feature a reversed value pattern of these physical quantities. However, because of a more homogeneous structure of the obtained diaphragms, high electrolysis product quality and current efficiency at a comparable power consumption it is recommended that SM-2 modifier be used.

Figure 1 *SEM microphotographs of the longitudinal sections cut parallel to the surface of the three-component diaphragm made from P-3-50 and P-4-20 asbestos grades in 1:1 proportion, with the addition of SM-2 modifier; a) cathode-adjacent zone, b) anode-adjacent zone.*

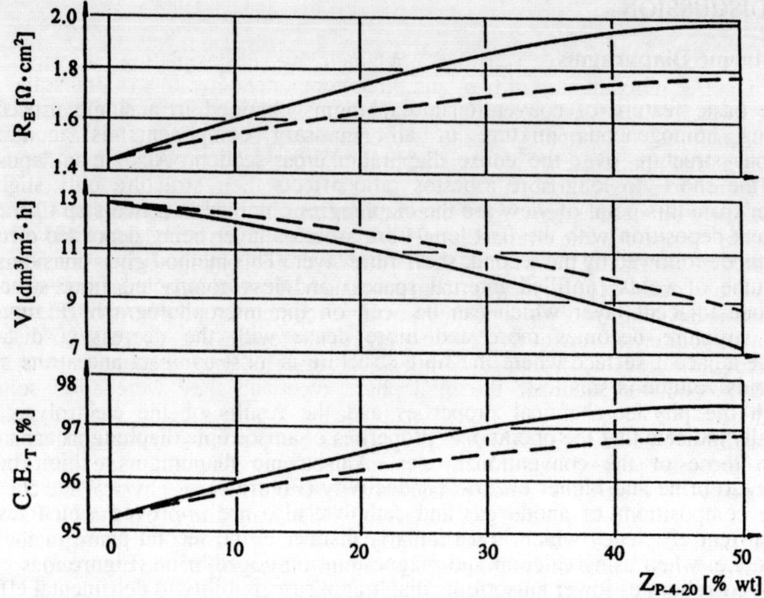

Figure 2 *The dependence of diaphragm permeability to brine (V), diaphragm effective ohmic resistance (R_E) and theoretical current efficiency ($C.E._T$) on the percentage of short-fibre asbestos ($Z_{P\text{-}4\text{-}20}$) in the mixture with the long-fibre, P-3-50 grade; solid line – single-layer diaphragm; dashed line – double-layer diphragm.*

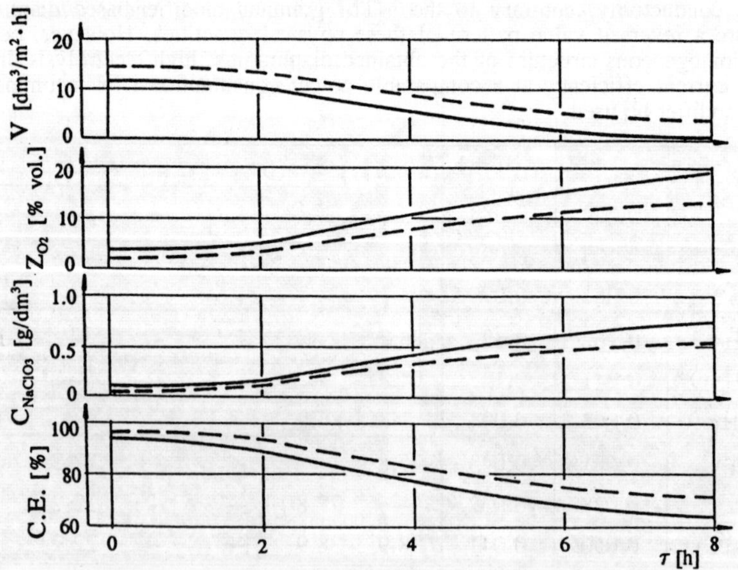

Figure 3 *Changes in diaphragm permeability to brine (V), oxygen content in anode gas (Z_{O_2}), chlorate concentration in catholyte (C_{NaClO_3}) and current efficiency (C.E.) during the test electrolysis (τ); solid line – single-layer diaphragm; dashed line – double-layer diaphragm.*

4.2 Activated Cathodes

In the case of active coatings obtained by electroplating it was found that the content of molybdenum in two- and three-component coatings is almost the same amounting to ca 25 % wt, whereas that of vanadium is about six times higher in three-component coatings as compared to the two-component ones. From this a conclusion can be drawn that the presence of molybdenum favours the process of co-deposition of vanadium in the alloy (Table 1).

Nickel coatings obtained by the chemical method at pH=3 contain ca 13.5 % wt of phosphorus, whereas those obtained at pH = 5 – 10 contain much less of this element, within 5 – 8 % wt, with the precipitation rate of these coatings being much higher. The process of cobalt coating deposition runs much more slowly than that of nickel coatings and the cobalt coatings deposited at pH=9 contain ca 13 % wt of phosphorus, whereas in an acid environment this process practically ceases.

Taking the results of phase composition analysis into account it was found that the alloys obtained were of the monophase type and they were solid solutions of molybdenum and vanadium in nickel, with the Ni-V alloys being well crystallised. Notwithstanding, Ni-Mo and Ni-Mo-V alloys were very poorly crystallised and their diffraction patterns showed features of amorphous materials with wide and strongly broadened peaks. These results indicate that molybdenum influences physical and chemical properties of the Ni-Mo-V coatings more strongly than vanadium. On the Ni-P coating diffraction patterns, apart from the reflections coming from the substrate, weak and strongly diffused maxima can be seen, being an indication of the amorphous nature of the coatings studied.

In the coatings obtained from alkaline solutions at pH 9 and 10 the presence of fine-crystalline nickel phosphide was found, whereas on the diffraction patterns of the coatings obtained by thermal method NiO reflections and weak reflections coming from platinum were observed.

Table 1 *Coefficients a and b in Tafel equation:* $\eta = a + b \cdot \log i$ *[mA/cm²], for the process of hydrogen evolution on alloy electrodes in 1 M NaOH solution.*

Coating	$-a$ [V]	$-b$ [V]	Coating constituents [% wt]: Ni	Co	Mo	V	Pt	P
Ni	0.248	0.129	100	–	–	–	–	–
Ni-Mo	0.079	0.089	75.3	–	24.7	–	–	–
Ni-V	0.108	0.098	99.8	–	–	0.2	–	–
Ni-Mo-V	0.061	0.079	74.2	–	24.6	1.2	–	–
Ni-P;pH=3	0.162	0.123	86.7	–	–	–	–	13.3
Ni-P;pH=9	0.241	0.127	94.3	–	–	–	–	5.7
Co-P;pH=9	0.063	0.094	–	86.9	–	–	–	13.1
Ni-Pt*)	0.051	0.043	77.8	–	–	–	22.2	–
Co-Pt*)	0.028	0.032	–	77.8	–	–	22.2	–
Ni-Co-Pt*)	0.050	0.041	38.9	38.9	–	–	22.2	–

**) - after cathodic reduction in 1 M NaOH at i = 500 mA/cm² current density, over 25 h period.*

The microscopic examinations revealed that the Ni-V coatings are characterised by the presence of large-size domed surface irregularities and a very dense microcrack network, whereas the Ni-Mo coatings have a relatively even surface but also a distinct network of cracks. The Ni-Mo-V coatings resembled the Ni-Mo coatings. The surface of the coatings deposited chemically from solutions in the pH = 3 – 9 range were relatively smooth and practically without cracks.

The morphology of thermal coatings differs distinctly from that of the remaining coating types and is strongly affected by the substrate structure which is an indication of a relatively small thickness of this type of coating. Scratches, craters and cracks are observed on their surfaces, which formed probably on solvent evaporation. The 25 h reduction process leaves a relatively high number of cracks and in some rather rare places - defects in the top layer of the coating (Figure 4).

The results of the double layer capacitance measurements showed that the Ni-V alloy coatings feature the highest development of active surface. These coatings have their capacitance minimum at a potential of about -0.65 V vs SCE being normally over 2000 $\mu F/cm^2$. These values are 120 – 150 times higher than the capacitance of the mercury electrode double layer, whose real surface area is equal to its geometric area and its capacitance is ca 17 $\mu F/cm^2$. The Ni-Mo and Ni-Mo-V coatings exhibit somewhat lower capacitances, which are 60 – 100 times higher than the capacitance of the mercury electrode, and show similar relationship between capacitance and potential with the minimum at ca -0.65 V vs SCE.

Nickel-phosphorus and cobalt-phosphorus coatings obtained by the chemical method have similar double layer capacitances. Capacitance minimum for these coatings is at a potential of ca -0.8 V vs SCE and ranges from 30 $\mu F/cm^2$ to 100 $\mu F/cm^2$ which corresponds to the 2 to 6-fold active surface development.

Double layer capacitance of the thermally obtained coatings decreases with the thermal treatment temperature decrease and is ca 100 $\mu F/cm^2$ in the case of 450 °C and ca 20 $\mu F/cm^2$ in the case of 300 °C. The use of the cathodic reduction process leads to a radical increase in active surface area. The capacitance of the electric double layer of the electrodes having Ni-Pt coating subjected to 25 hour reduction is ca 800 $\mu F/cm^2$ whereas that of the Co-Pt coated electrodes can reach as much as ca 2000 $\mu F/cm^2$ which corresponds to 50 up to 120 times increase in the active surface area. After reduction the capacitance minimum is at ca -0.55 V vs SCE.

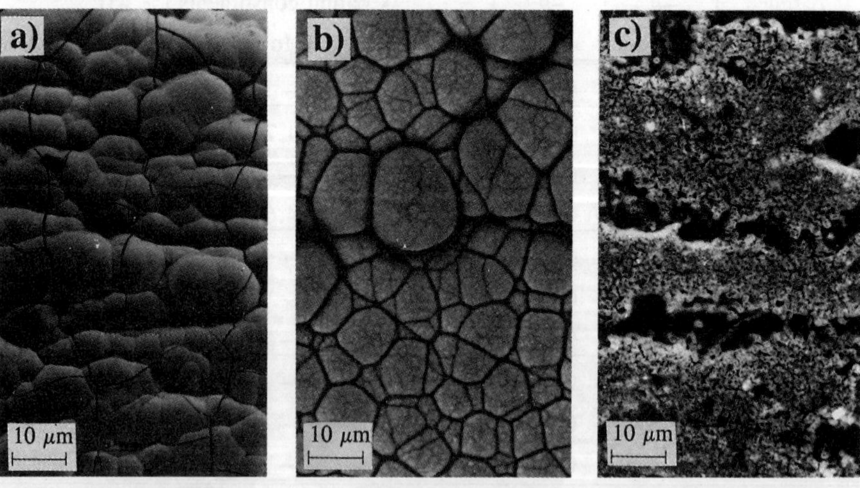

Figure 4 *SEM microphotographs of the coating surfaces: a) nickel-molybdenum-vanadium coatings obtained by electroplating, b) cobalt-phosphorus obtained using the chemical method, c) cobalt-platinum obtained in the process of thermal decomposition of corresponding salts.*

The Ni-Mo and Ni-Mo-V alloy coatings show similar hydrogen overpotential values. The *a* and *b* coefficients in the Tafel equation expressed in volts, within the range of current densities of 1 – 100 mA/cm^2, in alkaline solution are: -0.06 – -0.08 V and -0.08 – -0.09 V correspondingly. The Ni-V alloy coatings show slightly higher values of these coefficients. The results obtained are indicative of the favourable influence of both molybdenum and vanadium in a three-component alloy on its properties, which has then much better hydrogen evolution properties as compared with two-component alloys. Among them the nickel-molybdenum alloy most closely rsembles a three-component one. Similar values of Tafel equation coefficients in the case of Ni-Mo and Ni-Mo-V alloys corroborate the marked influence of the molybdenum and a weaker influence of the vanadium presence on the alloy properties.

A more pronounced influence of these alloying additives can be seen when analysing exchange current which is considered to be a measure of the electro-catalytic activity of a cathode material. The values of exchange current, i_o, determined at a potential of $E_{i=0}$, at which no resultant current is flowing, are: for a three-component alloy ca 6.0 mA/cm^2, for Ni-Mo alloy – ca 2.8 mA/cm^2 and for Ni-V – ca 1.8 mA/cm^2.

Amongst chemically obtained coatings, the cobalt coatings exhibit good electrochemical properties. Their *a* and *b* coefficients in alkaline solution are -0.06 V and -0.09 V correspondingly. The nickel coatings obtained by the same method show relatively high hydrogen overpotential, having their *a* and *b* coefficients in the same solution -0.16 – -0.24 V and -0.12 – -0.13 V correspondingly.

Amongst coatings obtained by thermal decomposition of corresponding salts, the cobalt-platinum coatings subjected to 25 h cathodic reduction have the lowest hydrogen overpotential. Their *a* and *b* coefficients are -0.03 V and -0.03 V correspondingly, which gives ca 0.06 V overpotential in alkaline solution, at i =10 mA/cm^2 current density. These values are better than those of three-component electroplated coatings. For the remaining coating types obtained by this method the overpotential is slightly higher and is ca 0.09 V. Exchange current values for these coatings are ranging within 0.8 – 0.9 mA/cm^2 and after a cathodic reduction - within 1.2 – 1.5 mA/cm^2.

4.3 Cathode-Diaphragm Systems

The laboratory tests with the use both of anisotropic diaphragms and activated cathodes confirmed good operational properties of the cathode-diaphragm system investigated. The terminal voltage of the experimental cell was reduced by ca 0.3 V, which corresponds to an 8 – 10 % drop in power consumption, when the cathodes activated by nickel micromesh, the latter covered with Co-Pt active coating and the asbestos-polymer anisotropic diaphragms based on long-fibre (P-3-50) and short-fibre (P-4-20) asbestos grades with SM-2 modifier added were used (Figure 5). Better results, corresponding to up to ca 15% reduction in specific power consumption, were achieved when a cobalt-platinum alloy-activated carbon fibre layer was laid on the cathode wire net directly beneath the anisotropic diaphragm having the same features as mentioned earlier, but obtained with the use of HALAR modifier - because of the diaphragm baking temperature. Owing to the lower susceptibility of anisotropic diaphragms to the adverse effect of impurities such as calcium and magnesium ions, a slower rate of permeability-to-brine drop is observed which promises longer cell repair-to-repair periods, with the technical and economic parameters of the electrolysis process maintained and even in some cases improved.

5 CONCLUSION

The laboratory scale investigations show unequivocally that a simultaneous use of anisotropic diaphragms and cathodes activated by depositing cobalt-platinum coatings leads to a considerable improvement of the parameters of the sodium chloride diaphragm electrolysis process. The industrial tests started recently aim at the verification of the concept of the cathode-diaphragm system presented in the paper.

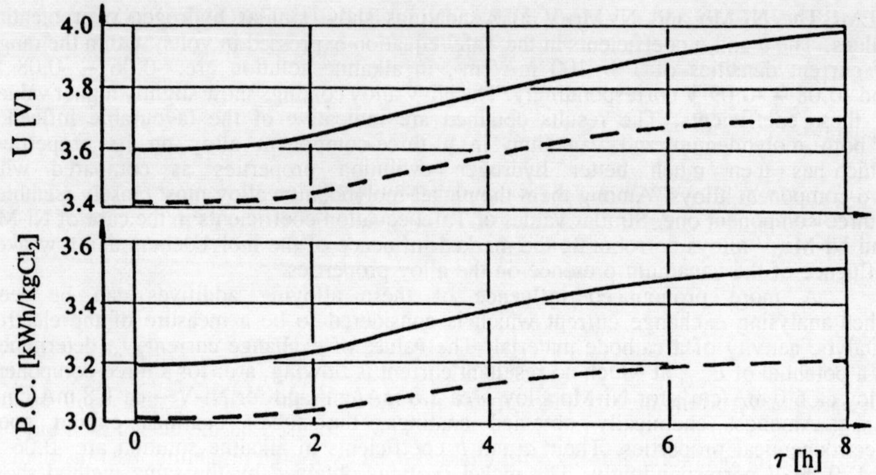

Figure 5 *Changes in cell terminal voltage (U) and electric energy consumption (P.C.) during the test electrolysis (τ); solid line – single-layer diaphragm, cathode without active coating; dashed line – double-layer diaphragm and cathode activated with nickel mesh with Co-Pt coating.*

6 ACKNOWLEDGEMENT

The study was financially supported by the Committee for Scientific Research (KBN) in Warsaw.

7 REFERENCES

1. G. Nawrat, M. Gonet, A. Małachowski and A. Korczyński, *Polish J. Appl. Chem.*, 1996, **40**, 309.
2. T. Buczek-Karczewska, A. Dukowicz, G. Nawrat, A. Małachowski and A. Korczyński, *Polish J. Appl. Chem.*, 1996, **40**, 129.
3. G. Nawrat, T. Buczek-Karczewska, M. Gonet and A. Korczyński, 'Modern Chlor-Alkali Technology', Ed. R.W. Curry, SCI, London, 1995, vol.6, ch.15, p.149.
4. G. Nawrat, T. Buczek-Karczewska, A. Małachowski and A. Korczyński, *Polish J. Appl. Chem.*, 1996, **40**, 135.
5. V.H. Thomas and E.J. Rudd, 'Modern Chlor-Alkali Technology', Ed. C. Jackson, Ellis Horwood Limited, Chichester, 1983, vol.2, ch.10, p.159.
6. M.H. Miles and M.A. Thomason, *J. Electrochem. Soc.*, 1976, **123**, 1459.
7. M.D. Archer, C.C. Corke and B.H. Harji, *Electrochim. Acta*, 1987, **32**, 13.
8. H. Isfort, 'Modern Chlor-Alkali Technology', Ed. C. Jackson, Ellis Horwood Limited, Chichester, 1983, vol.2, ch.9, p.132.
9. K.A. Poush, D.L. Caldwell, J.W. Van Zee and R.E. White, 'Modern Chlor-Alkali Technology', Ed. C. Jackson, Ellis Horwood Limited, Chichester, 1983, vol.2, ch.2, p.21.
10. R.E. White, J.S. Beckerdite and J.W. Van Zee, 'Electrochemical Cell Design', Ed. R.E. White, Plenum Press, New York, 1984, p.25.

11

MEMBRANE ELECTROLYZER OPERATING AT HIGH CURRENT DENSITY

Luciano Iacopetti

Electrolyzer Design/Direction of Research & Technology
DE NORA S.p.A.
20134 Milan - Italy

1 INTRODUCTION

The new high current density electrolyzer illustrated in this paper is the result of the development program carried out by De Nora S.p.A., focused on the design of a bipolar electrolyzer for operation at 6-8 kA/m^2 characterized by a high reliability, best membrane performance and attractive energy consumptions (patent pending).

De Nora has a remarkable experience in the electrochemical field, gathered over 70 years of activity in the market and encompassed in a proprietary know-how for the construction of plants, electrolyzers and relevant special components, in particular electrodes and coatings.

De Nora started the development of its proprietary membrane electrolyzers in the seventies and for over twelve years has been commercialising the well-known *"DD"* electrolyzers developed in the frame of a joint-venture with Dow Chemical. For a long time De Nora has been the only supplier of both monopolar and bipolar electrolyzers based on the same technology.

This long-term experience in the engineering and manufacturing together with the feedback information from plant sites have been instrumental in the development of a new generation electrolyzer with first-rate operation characteristics.

The family of the new generation electrolyzers is designated as

DN x x x

the last three digits corresponding to the area of active surface for each element, expressed as square decimetres. The standard reference is 350 dm^2.

2 HIGH CURRENT DENSITY - MAIN REQUIREMENTS

Four operating ranges have been defined, as reported hereinafter:

1	Low current density	< 3.5	kA/m^2
2	Conventional range	3.5 - 4.5	kA/m^2
3	High current density	4.5 - 6.0	kA/m^2
4	Peak current density	6.0 - 8.0	kA/m^2

The high current density field has been split in view of the characteristics of the presently available ion exchange membranes, which guarantee the necessary reliability up to 6.0 kA/m^2, whereas performances have to be still demonstrated in the peak current density range. The design point of the new electrolyzer has been defined as 8.0 kA/m^2 to permit the operation of any low ohmic drop membrane of future conception. The operating characteristics of any high current density electrolyzer are substantially connected to an appropriate performance of the ion exchange membrane. The main influencing factors are reported in Table 1.

The factors reported in Table 1 directly influence the operating characteristics of the electrolyzer, as shown in Table 2.

Table 1 *Main Requirements for High Current Density Operation*

1	Efficient mixing of the bulk electrolytes both in the anodic and cathodic compartments	=	Temperature uniformity across the membrane surface	=	± 3 °C
2	Efficient mixing of the bulk electrolytes both in the anodic and cathodic compartments	=	Uniformity of brine concentration in contact with the membrane	=	210 ± 5 gpl NaCl
3	Efficient mixing of the bulk electrolytes both in the anodic and cathodic compartments	=	Uniformity of caustic soda concentration in contact with the membrane	=	32 ± 0.2 % NaOH
4	Optimum pitch and sizing of the electrical contacts	=	Even macro-distribution of the current density on the membrane	=	+0.1/-0.2 %
5	High mass transfer between membrane and bulk brine	=	Limited gradient of the brine overdepletion close to the membrane	=	≤ 5 gpl NaCl
6	High mass transfer between membrane and bulk caustic soda	=	Limited gradient of caustic soda concentration close to the membrane	=	≤ 0.2 % NaOH
7	Optimum geometry of the electrodic structures	=	Uniform microdistribution of current on the membrane and low bubble effect	=	± 0.05 %
8	Reduction of the current paths through the electrolytes	=	Reduction of ohmic drops and energy consumption	=	Cathodic zero-gap
9	Stable discharge of the mixed gas-liquid flow (low pressure fluctuations)	=	Reduction of mechanical stresses on the membrane (negligible abrasion)	=	< 7 mbar < 0.5 Hz
10	Gas pockets elimination (mainly anodic)	=	Prevention of polymer embrittlement by salt precipitation	=	No pockets

Table 2 *Key Factors Effect on High Current Density Operation*

Factor	*Main effect*	*Secondary effect*
1,2,5	Durability of membrane performances with time	Preservation of appropriate mechanical characteristics of the polymer
3,6	Durability of membrane performances with time	• Cell voltage reduction • Polymer durability
4,7	Cell voltage reduction	Maintenance of current efficiency with time
8	Cell voltage reduction	
9	Prevention of mechanical decay of the polymer	
10	Prevention of mechanical decay of the polymer	Durability of membrane performances with time

3 MAIN CHARACTERISTICS OF THE DE NORA MEMBRANE ELECTROLYZER

The main characteristics of the DN type bipolar electrolyzer, extrapolated from long term operation of a large set of laboratory cells, are:

- Design high current density — *8.0 kA/m²*
- Low cell voltage — *3.28 V at 6 kA/m²*
- Stable current efficiency — *$\geqslant$ 93% after 3 years*
- High active area for each bipolar element — *up to 3.5 m²*
- High number of bipolar elements for each stack — *up to 90*
- Possibility of operating under pressure — *up to 0.5 barG*
- Possibility of operating at high temperature — *up to 100 °C*
- O_2 control by acid addition — *$\leqslant$ 0.5 % O_2 in Cl_2*
- Long operating lifetime of the membranes — *$\geqslant$ 4 years*
- Chemically inert gaskets — *$\geqslant$ 5 years*

An exploded view of the DN350-type electrolyzer is shown in Figure 1. A discussion of the main design details is given in the following sections.

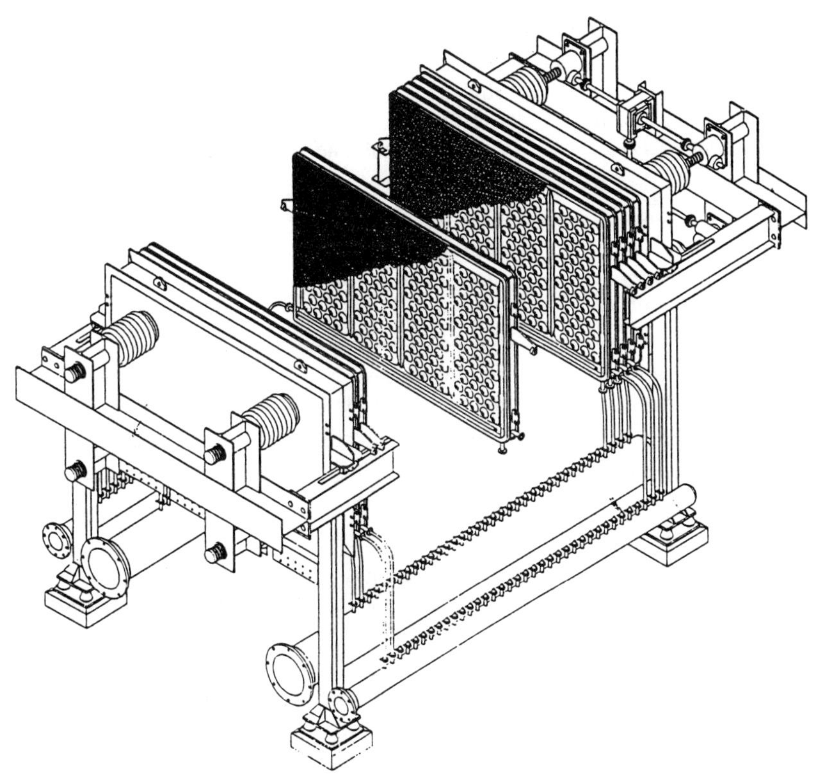

Figure 1 *DN350 type Electrolyzer - Exploded View*

4 MEETING THE MAIN REQUIREMENTS

Table 3 summarises the main design concepts adopted in the new electrolyzer.

Table 3 *Design Concepts for High Current Density Operation*

1	*Natural recirculation inside both anode and cathode compartments*	⇒	High ratio between the inlet flow-rate and the recirculation flow rate	⇒	Minimum thermal and material gradient inside the anode and cathode compartments
2	*Anode geometry with enhanced gas release*	⇒	Enhanced local circulation of brine with best release of the chlorine bubbles	⇒	Minimum gradient between bulk and anode/ membrane interface
3	*Cathode zero-gap configuration based on elastically deformable structures*	⇒	Minimum ohmic drop in the electrolytes	⇒	Low cell voltage and mechanical support for the membrane
4	*Cathode structure with low membrane blinding effect*	⇒	High exchange of caustic soda at the cathode-membrane interface	⇒	Minimum gradient between caustic soda bulk and cathode/membrane interface and negligible bubble effect
5	*Quincunx arrangement of the electrical connections*	⇒	Homogeneous macrodistribution of current	⇒	Minimum macro-gradients of current density on the active surface
6	*Suitable void-ratio of the electrode structures*	⇒	Homogeneous microdistribution of current	⇒	Minimum microgradients of current density on the active surface
7	*Stand-pipe for gas-liquid downflow discharge mode*		Low amplitude and frequency of the pressure pulsation	⇒	Prevention of mechanical damages to the membrane
8	*Baffle on the top of the compartments*	⇒	No stagnant gas pockets on the top of the elements	⇒	Prevention of salt precipitation inside the membrane (embrittlement)
9	*Dimensionally stable and corrosion resistant gasketing*	⇒	Long operating life of both the anodic and cathodic gaskets (five years minimum)	⇒	Reduced stresses on the periphery of the membranes

5 LOW CELL VOLTAGE

5.1 Energy Losses Due to the Structure

The DN type bipolar element is sketched in Figure 2. Its structure, which derives from the conventional DD type, exhibits a remarkably low electrical resistance.

The electric current directly flows from the anode to the cathode surface through an array of welded electrical connectors made of highly conductive metals (e.g. steel or copper). These connectors are interposed between the titanium anodic wall and the nickel cathodic wall to which they are welded following a specifically developed procedure. Before assembling, both the anodic and cathodic walls are cold-pressed to create a pattern of bulges.

The weld spots are located within the top area of the bulges, which also act as electrode supports. That is, both the anode and cathode are directly welded on the same spot where the anode and cathode walls are welded to the electric connectors. No stand-off elements are interposed between anode, cathode and the respective walls.

The quincunx arrangement and density, that is the pitch, of connectors (Figure 3) provide for the most homogeneous macrodistribution of current on the electrode surface. The pitch between electrode periphery and the closest electric connectors is purposely increased to permit operation at a lower current density (about 20% less than the nominal one) on the peripheral area of the membrane.

5.2 Zero Gap Cathodic Configuration

The zero-gap configuration adopted for the DN type electrolyzer, alternative to the conventional structure, avoids gas trapping and increases the mass transfer towards the caustic soda bulk at the highest current densities.

The structure shown in Figure 4 is based on a mesh, cold pressed to create an array of bumps, which may be easily compressed without loosing their elasticity. Hence, the cathode mesh can adapt to the membrane profile exerting very low pressures which are not dangerous for the polymer integrity. The cathodic mesh is welded in correspondence of the current connectors.

The main advantages offered by the DN-type cathodic zero-gap configuration are illustrated in Table 4.

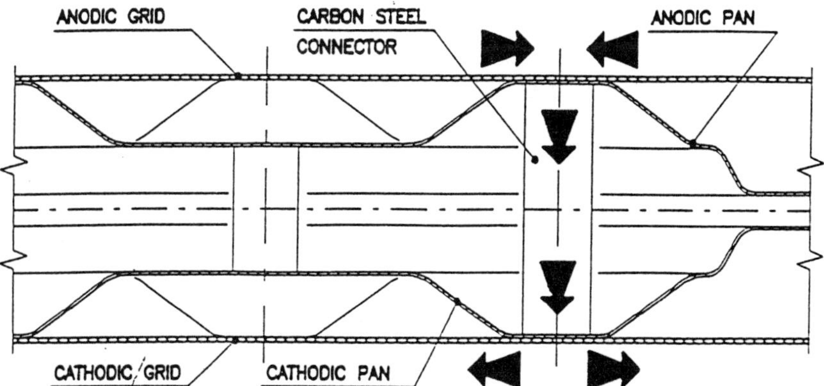

Figure 2 *DN350 type - Bipolar Structure*

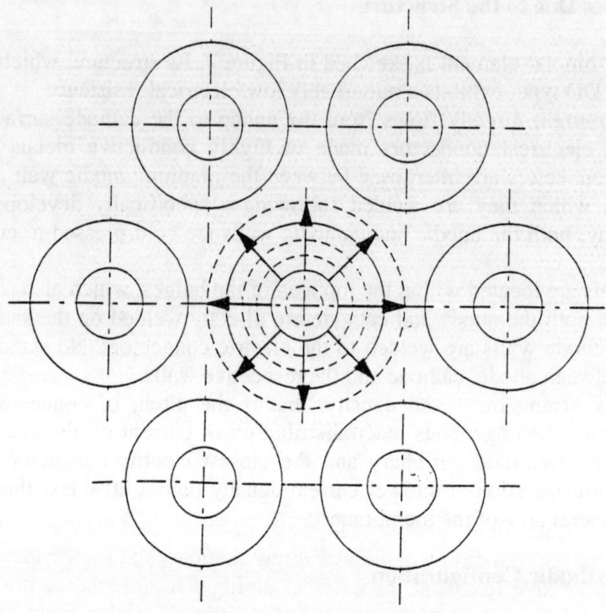

Figure 3 *Electrical Macrodistribution from Hexagonal Centered Array of Connectors*

Table 4 *Main Advantages of the DN-type Cathodic Zero-Gap Configuration*

1	Mechanically simple structure	*Forgiving mechanical tolerances*
2	Welded electrical contacts	*Time-independent current distribution*
3	Geometrically open structure of the electrode meshes	*Enhanced mass transfer towards the bulk electrolytes*
4	Easy gas release	*Reduced bubble effect*
5	Mechanical support to the membrane	*Negligible swelling effect*
6	Replaceable structure	*Cheap and quick recoating procedure*

6 HYDRAULICS

6.1 Internal Recirculation

The natural internal recirculation inside the anode and cathode compartments derives from the design adopted by De Nora for the conventional DD type electrolyzers (natural external recirculation). The decision to shift from external to internal mode is a consequence of the expected operational current densities which require specific

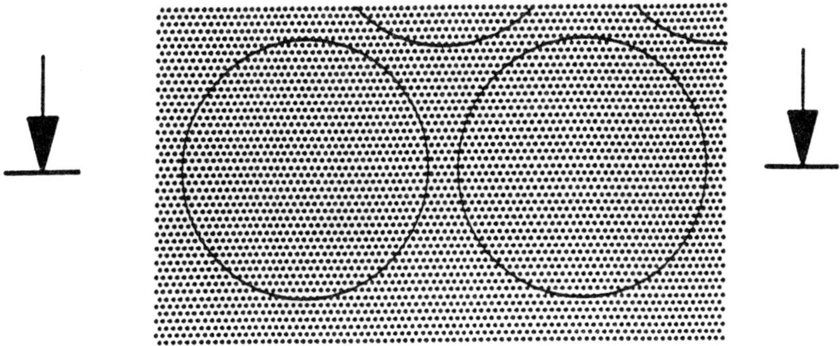

Figure 4 *Zero Gap Cathodic Structure*

recirculation rates substantially higher than the traditional ones. Under these conditions, the external natural recirculation would be severely bottlenecked by the insufficient cross-section and the excessive length of the external piping. The internal natural recirculation applied in the new DN type electrolyzer, shown in Figure 5, is based on an upper baffle (Figure 6) which accelerates the flow of the gas-liquid mixed phase, induces a forced gas separation and allows for the collection of the separated gas and liquid in the backspaces, between the baffle itself and the compartment backwall.

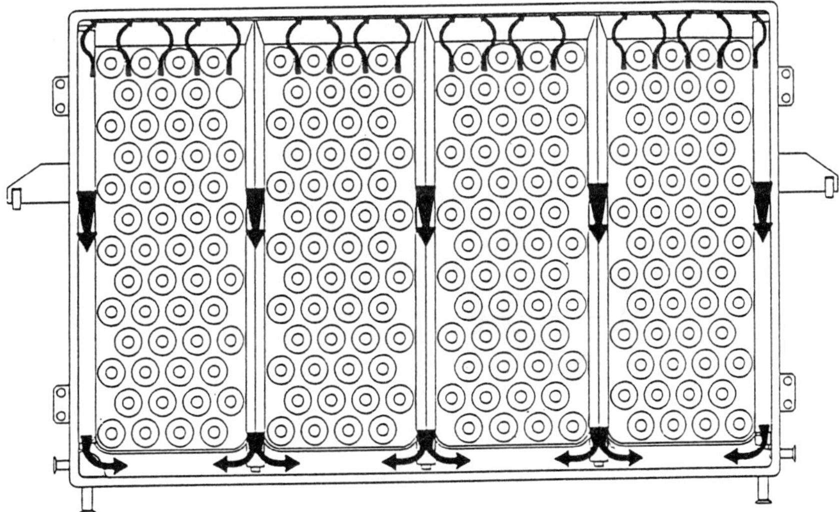

Figure 5 *DN350 Electrolyzer - Internal Recirculation System*

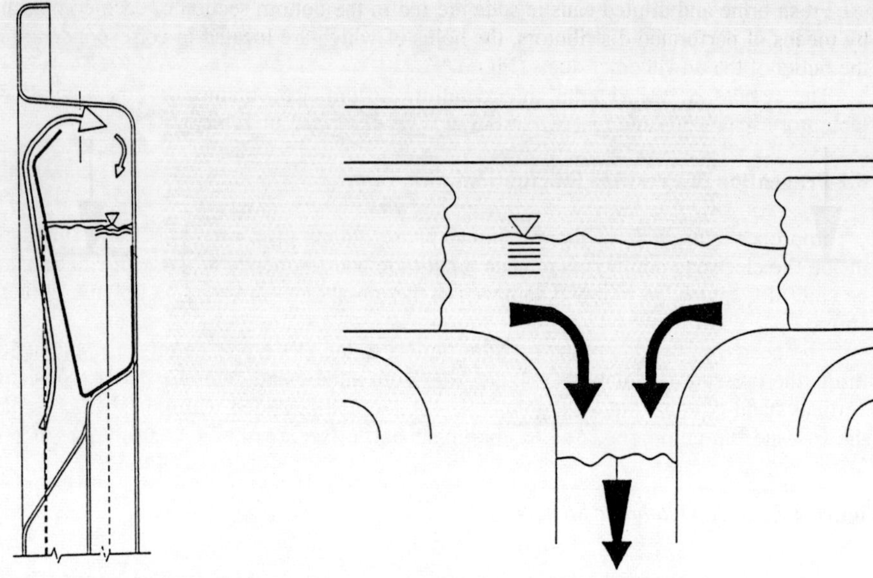

Figure 6 *Degassing Upper Baffle* **Figure 7** *Downcoming Ducts Inlet*

The gas-free liquid is channelled to the bottom of the compartment through downcomer ducts (Figure 7). The density differential between the mixed gas in the compartment space and the gas-free liquid in the downcomer ducts promotes a very active internal recirculation of the electrolyte.

Distributors positioned at the outlet of the downcomer ducts disperse the recirculated liquid over the whole active surface. Further, the quincunx arrangement of the bulges, already described in Section 5.1, further contributes to enhancing cross-mixing of the electrolyte (Figure 8).

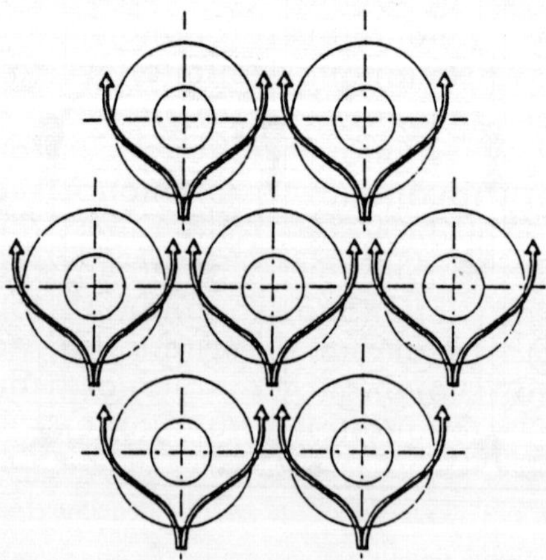

Figure 8 *Cross-Mixing of the Electrolyte Enhanced by the Centered Hexagonal Array of Bulges*

Fresh brine and diluted caustic soda are fed in the bottom section of each compartment by means of perforated distributors, the holes of which are located in correspondence with the outlet of the downcomer ducts (Figure 9).

The results of the internal recirculation system are obtained by the operation of laboratory scale cells and real-size mock-ups, as described in Table 5.

6.2 Prevention of Pressure Fluctuations

Improper hydraulics of the gas-liquid phase outlets give rise to pressure fluctuations inside the electrode compartment. The amplitude and frequency of these fluctuations must be carefully controlled to avoid damages to the membrane (erosion and polymer structural stress).

The design of the gas-liquid mixture outlets in the DN type electrolyzers satisfactorily limits the pressure fluctuations (Figure 10). Both anodic and cathodic outlets consist of a vertical stand-pipe having a suitable cross-section and provided with inlets which separate the gas and the liquid degassed by the upper baffle (see Section 4.4), in the absence of a

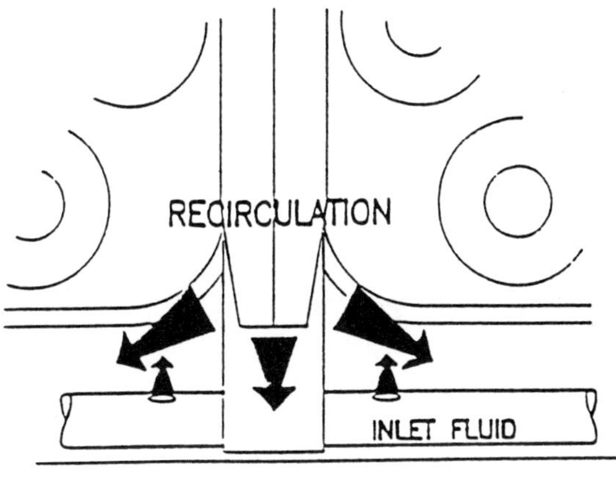

Figure 9 *Downcoming Ducts Outlet*

Table 5 *Internal Recirculation System - Operating Data*

Upper Gas Separation Baffle + Down-comer Ducts + Fresh Brine Distributor		
1	Specific internal recycle flow rate of the anolyte at 8.0 kA/m^2	**2.8 m^3/h/m^2**
2	Specific internal recycle flow rate of the catholyte at 8.0 kA/m^2	**2.4 m^3/h/m^2**
3	Specific internal recycle flow rate of the anolyte at 5.0 kA/m^2	**2.3 m^3/h/m^2**
4	Specific internal recycle flow rate of the catholyte at 5.0 kA/m^2	**2.0 m^3/h/m^2**
5	Recycle-fresh brine flow rates ratio at 8.0 kA/m^2	**20**
6	Recycle-fresh brine flow rates ratio at 5.0 kA/m^2	**27**
7	Maximum/minimum deviation of the anolyte concentration	**± 3 gpl**
8	Maximum/minimum deviation of the catholyte concentration	**± 0.2 %**
9	Maximum/minimum temperature deviation	**+ 1 / -2 °C**

counter-pressure. This flow-pattern prevents any pressure instability and the amplitude of the residual pressure fluctuation is comprised between 2.5 and 7 mbar in an operating range of 4.0-8.0 kA/m².

7 ENERGY CONSUMPTION

The performances of the DN type electrolyzer are reported in the diagram of Figure 11.

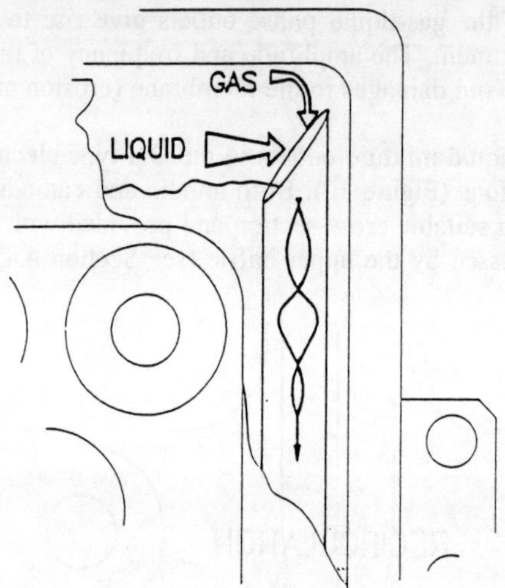

Figure 10 *Vertical Discharge Stand Pipe Arrangement*

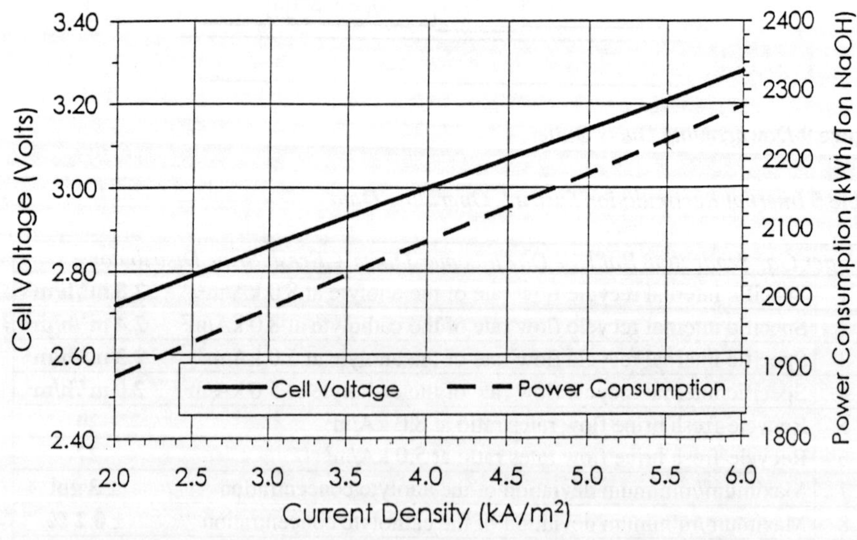

Figure 11 *Cell Voltage and Power Consumption for DN350 Electrolyzer Type*

12

HIGH PERFORMANCE OPERATION WITH FLEMION MEMBRANES AND THE AZEC-B1 ELECTROLYZER

M. Nakao*, T. Shimohira** and Y. Takechi*

*Chemical Engineering Division, Asahi Glass Co., Ltd.,
2-25-14, Kameido, Koutou-ku, Tokyo 136, Japan

**Chemical Research Laboratory, Asahi Glass Co., Ltd.
1150, Hazawa-cho, Asahi-ku, Yokohama 221, Japan

1 INTRODUCTION

During last 20 years Asahi Glass Co., Ltd. (AGC) has made an intensive effort to realize a high performance operation as a technology leader both of the membrane and the electrolyzer. AGC has supplied Flemion membranes to more than 100 users all over the world, and licensed its AZEC process to more than 50 chlorine producers whose capacity exceeds 3,500,000 MT/year (caustic base).

The target of all the chlorine producers is to pursue the lowest operation cost in consideration of their own given conditions, aiming at the following operations;

(1) Low energy consumption
(2) Reliable operation with long life of cell components
(3) Easy and safe operation
(4) Flexible operation for varied conditions

Recently, in addition, high current density operation, for example at 5 kA/m^2, has been highlighted as an operational target. Although the high current density operation was recognized to be an attractive method to reduce the investment cost, it has not been widely adopted due to its technical difficulties. Generally in the high current density operation, membrane performance does not show its good performance obtained at lower current density, and cell operation becomes more difficult because the increasing current requires special cell design to achieve a homogeneous electrolyte distribution in the cell and the reliable discharging device, etc. AGC has made persistent and intensive efforts to overcome these technical difficulties expected at high current density, and has developed an efficient system with low voltage membranes and a new bipolar electrolyzer, AZEC-B1.

This paper focuses on the operation at high current density, and describes the fundamental properties of F893, a typical low voltage membrane, and the features of AZEC-B1, a new bipolar electrolyzer.

2 MEMBRANE PERFORMANCE AT HIGH CURRENT DENSITY

High current density operation requires the low voltage membrane as a prime condition, because in the case of the high voltage membranes the cell temperature rises beyond the upper limit of the appropriate range due to a high ohmic resistance, furthermore, it shows a faster drop in current efficiency with a lapse of time. Taking these into account, Flemion F893, a typical low voltage membrane, was examined to clarify the behavior of the membrane performance at high current density.

2.1 Fundamental Properties of the Membrane

The electrolytic properties of the membrane at higher current density are changed due to an increase in the transport of ions and electro-osmotic water, which determine the water content inside the membrane.

2.1.1 Ohmic Resistance of the Membrane. Figure 1 shows the ohmic resistance of the membrane, which means the net voltage attributable to the membrane, for different current densities. The ohmic resistance is not proportional to the current density, and it shows a convex curve at high current density. From this figure it is understood that the profile of the water content inside the membrane is changed with an increase in current density, and the average water content becomes larger at higher current density.

2.1.2 Effect of Caustic Strength on CE. The current efficiency of the membrane is greatly influenced by caustic strength, and the peak of the current efficiency is observed at the optimum caustic strength. If the caustic strength is lower or higher than the optimum value, the membrane state becomes more swollen or dehydrated, that lowers the current efficiency. Figure 2 shows the behavior of the current efficiency by changing the caustic strength under different current densities. With an increase in current density from 1 kA/m^2 to 6 kA/m^2, the peak of the current efficiency gradually decreases from 97.5% to 96%, and the optimum concentration to give the peak value shifts to the lower side. This figure also indicates that the high current density operation at more than 4 kA/m^2 brings a considerable decrease in current efficiency at the higher side of the caustic strength.

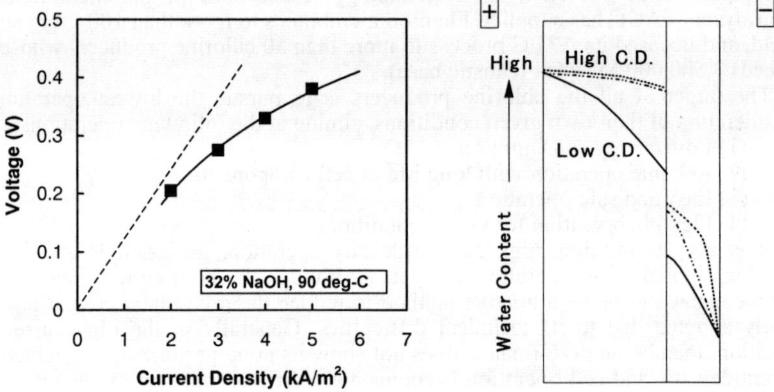

Figure 1 *Dependence of Ohmic Resistance of the Membrane on Current Density*

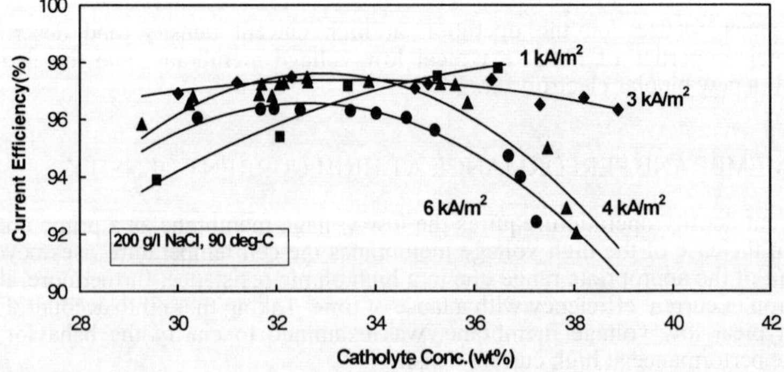

Figure 2 *Effect of Caustic Strength on Current Efficiency (Membrane: F893)*

The long run test under 6 kA/m^2 has been done by laboratory cells to clarify the difference between the operations of 32% NaOH and 35% NaOH. The results of the current efficiency are shown in Figure 3. The current efficiency obtained at 32% NaOH shows very stable, while that obtained at 35% NaOH shows a gradual decrease. The difference between the two operations expands from 0.5% at the initial stage to 2% after 4 months of operation.

2.1.3 Effect of Temperature on CE. The current efficiency is also influenced by operation temperature. Figure 4 depicts the results obtained by changing the cell temperature under different current densities. As the current density increases, the optimum temperature to give the peak of current efficiency shifts to the higher side. This figure indicates that the lower temperature should be avoided for the higher current density operation.

2.2 Anion Transport Through the Membrane

Anions, such as Cl, ClO_3 and SO_4, migrate through the membrane, and contaminate the caustic soda produced in the cathode compartment. It is known that the content of Cl and ClO_3 in the caustic soda is a commercially important item to be considered. Anion transport is governed by the following driving forces: electric field, diffusion and electro-osmotic flow. The influence of each driving force differs with each ion species. In the case of Cl, the dominant force is electric field, while electro-osmotic flow dominates the transport of SO_4. Since the current density influences both the intensity of electric field and the flux of electro-osmotic flow, the transport rate of anions is changed with an increase in current density. Figure 5 shows the concentration of Cl, and

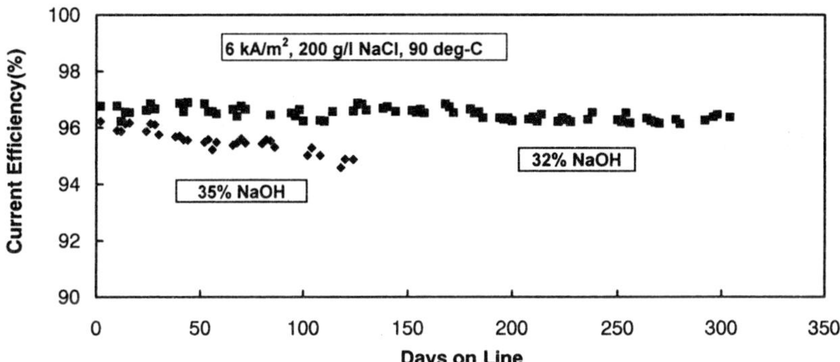

Figure 3 *Long Run Test under 6 kA/m^2 (Membrane: F893)*

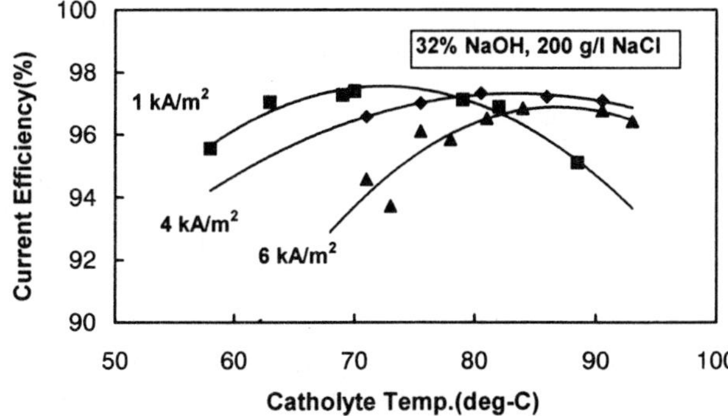

Figure 4 *Effect of Temperature on Current Efficiency (Membrane: F893)*

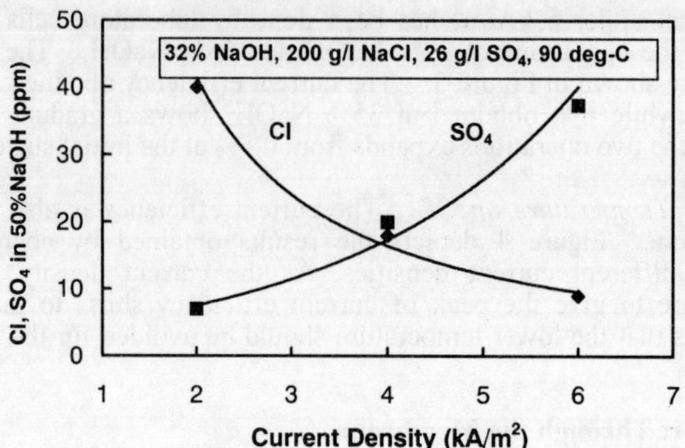

Figure 5 *Transport Rate of Cl, SO_4 vs Current Density (Membrane: F893)*

SO_4 in the caustic soda obtained for different current densities, which is related to the transport rate of the anions. The transport of Cl decreases with an increase in current density, but that of SO_4 shows a strange behavior which increases with an increase of current density. This data indicates that an additional specification of SO_4 in the brine is sometimes necessary for the consideration of the quality of caustic soda.

2.3 Effect of Brine Impurities on Current Efficiency

Extensive studies have been done to explain the effect of various types of brine impurities on the performance of the membranes, however, most of the data were obtained at normal current density of 3 kA/m^2. As described in section 2.1, the membrane properties are changed at higher current density; consequently, the different behavior of the brine impurities is expected. Taking typical impurities which lower the current efficiency, such as Ca, SO_4 and iodine, their effect on current density is discussed below.

2.3.1 Recovery Characteristic after Upset of High Ca Content. Calcium is a typical impurity in the feed brine which lowers the current efficiency; and the brine purification method to control Ca content at the "ppb" level has been established as a commercial process. However, in the actual operation, it is not the rare case that a high amount of Ca flows into the brine system due to a mis-operation, and lowers the current efficiency temporarily. Therefore, from the practical point of view it is important to know the recovery characteristics of the membrane after an upset of high Ca in feed brine. Figure 6 shows the behavior of the current efficiency after adding 2 ppm of Ca in feed brine at different current densities. Although the amount of Ca input to the cell was the same, the current efficiency shows different behavior, depending on the current density. With an increase in current density the decrease of the current efficiency becomes larger, and the recovery becomes smaller. The operation at high current density, consequently, requires a more careful control of the brine system.

2.3.2 Effect of Sulfate. Sulfate originates in the raw salt, and its content in feed brine is usually controlled at the level of 4-6 g/l SO_4 by adding barium or calcium salts in the primary brine purification. Most of the sulfate ions are transported through the membrane as described in section 2.2, and a small portion of them make a precipitate with sodium ions inside the membrane at the cathode side. Since a large amount of this precipitate lowers the current efficiency, an accelerated test to examine the effect on current density was done by adding a high level of sulfate continuously into the cell. Figure 7 and Figure 8 are the results obtained by adding 26 g/l of SO_4 for both the 35% NaOH and 32% NaOH operations. Under the condition of 32% NaOH there is no conspicuous difference among the data obtained for different current densities, which show a stable performance with a lapse of time. However, 35% NaOH operation gives different

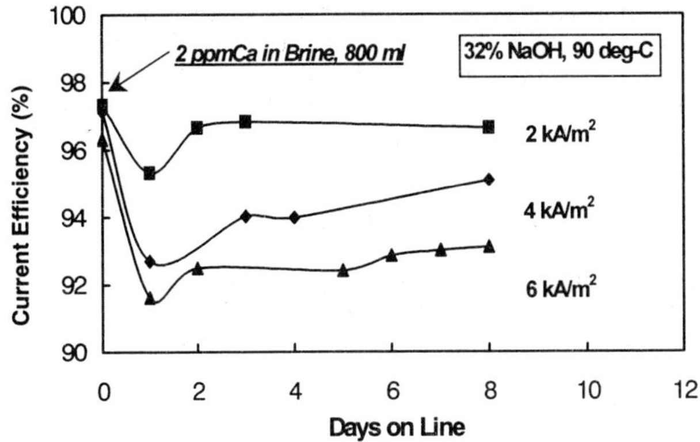

Figure 6 *Recovery Characteristics after Upset of High Ca Content*

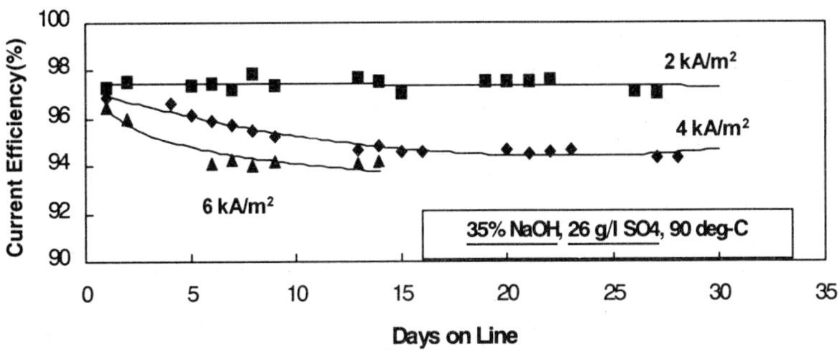

Figure 7 *Effect of Sulfate in Brine on Current Efficiency at 35%NaOH*

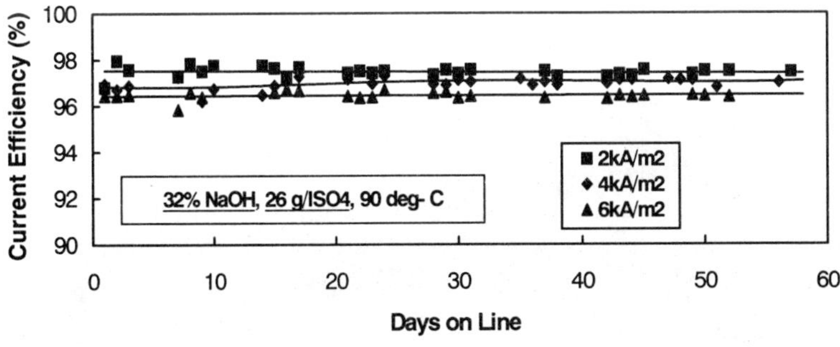

Figure 8 *Effect of Sulfate in Brine on Current Efficiency at 32%NaOH*

results, which show a rapid decrease in current efficiency at higher current density.

2.3.3 Effect of Iodine. Iodine, which exists in the feed brine in the range of 0.1 ppm (solar salt) to 0.5 ppm (well brine), negatively affects the current efficiency. A small amount of iodine transported into the membrane makes a precipitate inside the membrane on the cathode side in the presence of barium, and lowers the current efficiency. Although the whole aspect of this phenomenon has not been clarified, the behavior of the decrease in current efficiency due to barium/iodine contamination differs with membrane types and with the operating conditions. Figure 9 shows the results of an accelerated test, which is operated at 80 deg-C, 32% NaOH, with the addition of 10 ppm iodine and 1 ppm barium in the feed brine. The operation at higher current density shows a faster decrease in current efficiency, and within a few days it decreases to 92 % in the case of 6 kA/m^2. These data also reflect the difficulty of the operation at high current density.

3 AZEC-B1, A NEW EFFICIENT BI-POLAR ELECTROLYZER

The AZEC-B1 is a new filter-press type cell having the active electrolysis area of 2.88 m^2 per one sheet of membrane. The outward view of AZEC-B1 is shown in Figure10. AZEC-B1 is a well-designed and reliable cell to fulfill the various requirements, including the proper use of membranes which knowledge has been accumulated by AGC through much world-wide experience.

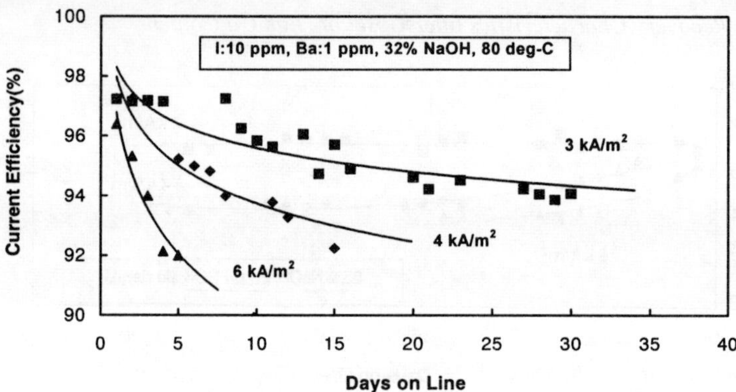

Figure 9 *Effect of Iodine in Brine on Current Efficiency (Membrane: F893)*

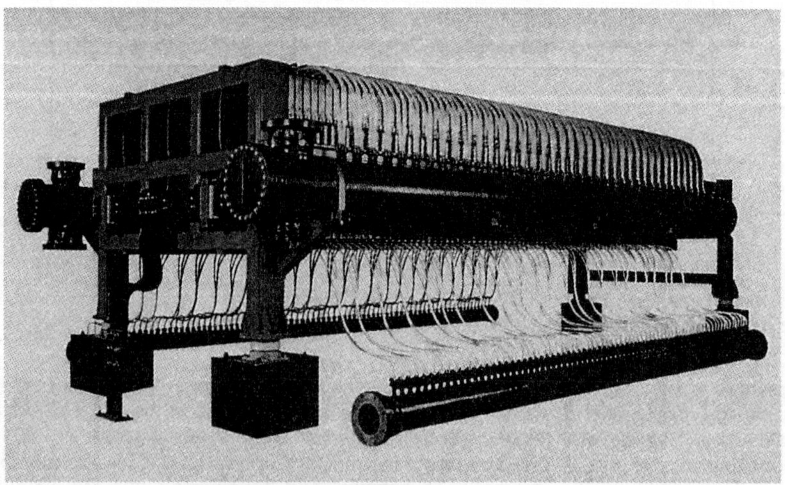

Figure 10 *Outward View of AZEC-B1 Electrolyzer*

3.1 Structural Characteristics of AZEC-B1

AZEC-B1 has a rigid structure composed of a titanium pan for the anode compartment and a nickel pan for the cathode one. Both pans are welded at a back-plate, and the rims of the pans, which form the flange parts for sealing purposes, are reinforced with square pipes. Numbers of conductive ribs having a special shape are welded onto the back-plate of both the anode and cathode compartments. This simple and rigid structure meets easy fabrication, and it realizes a good surface flatness of the anode and cathode meshes which ensures an efficient and stable operation for a long term. The details of the structure are described below.

3.1.1 "M" Shape Ribs. The functions of the ribs of the electrolyzer are firstly an electrical conductor to distribute the current uniformly to the electrode mesh, and secondly a structural support to give the electrolyzer sufficient rigidity. AZEC-B1 adopts a special rib, named the "M" shaped rib, having a cross sectional shape of the "M" character as shown in Figure 11. The "M" type rib is formed from the Ti or Ni metal sheet, and the feet of the rib are welded on the back-plate while the shoulder parts are welded onto the electrode meshes. The inside of the "M" type rib has a hollow space which works as a down-comer for the electrolyte circulation.

3.1.2 "Dam" Type Overflow System. At the upper place of the cell compartment which is isolated from the electrolysis section, a special space, named "Dam" type overflow, is installed. As shown in Figure 12 the Dam type overflow provides a narrow space at the backside of the flange near the back-plate where the discharged gas and electrolyte are rising as a foam. The foam is easily disengaged through the narrow space and the disengaged electrolyte overflows the dam and goes into a trough. Both the gas and the electrolyte are transported separately within the horizontal trough and discharged to the outlet nozzle.

3.2 Advantages of AZEC-B1

The above mentioned structural characteristics provide many advantages both in performance and reliability; that is, low power consumption, good and reliable current distribution, homogeneous electrolyte concentration, smooth discharge of gas and electrolytes, no electrical corrosion, etc. Details of the advantages are described below.

3.2.1 Homogeneous Electrolyte Distribution. Uniform concentration distribution of the electrolyte is the most important criterion for cell design in order to achieve good membrane performance, and it is accomplished in AZEC-B1 by sufficient internal

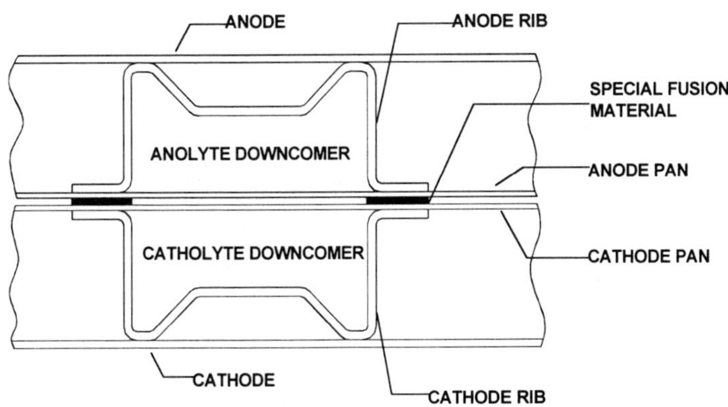

Figure 11 *"M" Shaped Rib and Bi-Polar Plate (AZEC-B1)*

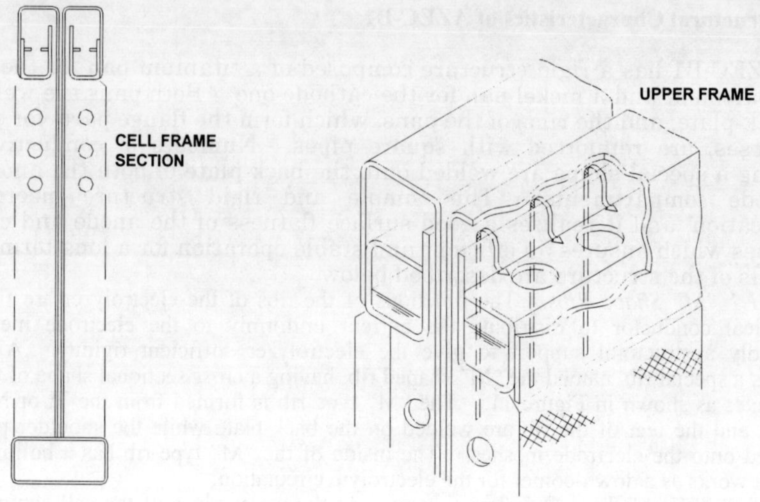

Figure 12 *Dam Type Overflow (AZEC-B1)*

circulation. As shown in Figure 13 twelve down-comers having a shape of the "M" character are installed in the cell in both anolyte and catholyte sides, and they bring a large amount of natural circulation. At 5 kA/m^2, the internal anolyte circulation ratio, for example, is approximately five times the feed brine rate. The homogeneous anolyte distribution is realized as shown in Figure 14, and it contributes to a stable operation for a long time. Figure 14 also indicates that the sufficient anolyte circulation ensures a reliable acidification process with the addition of hydrochloric acid to the brine for reducing the oxygen content in chlorine. In the AZEC-B1 electrolyzer a high purity chlorine gas (0.5% oxygen content) can be achieved by a well designed anode under reasonable anolyte pH condition, which does not cause membrane damage and titanium corrosion.

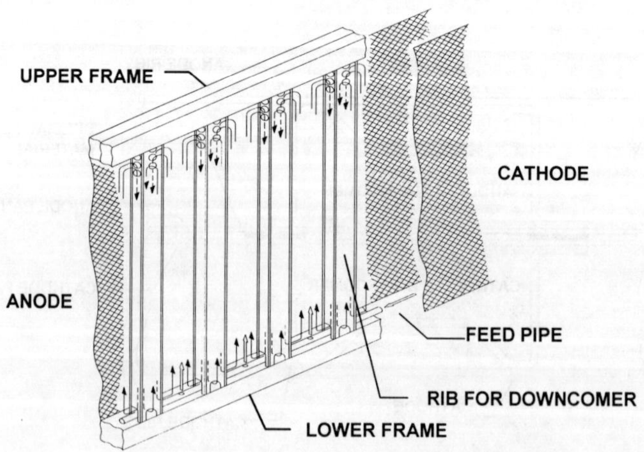

Figure 13 *Internal Electrolyte Circulation (AZEC-B1)*

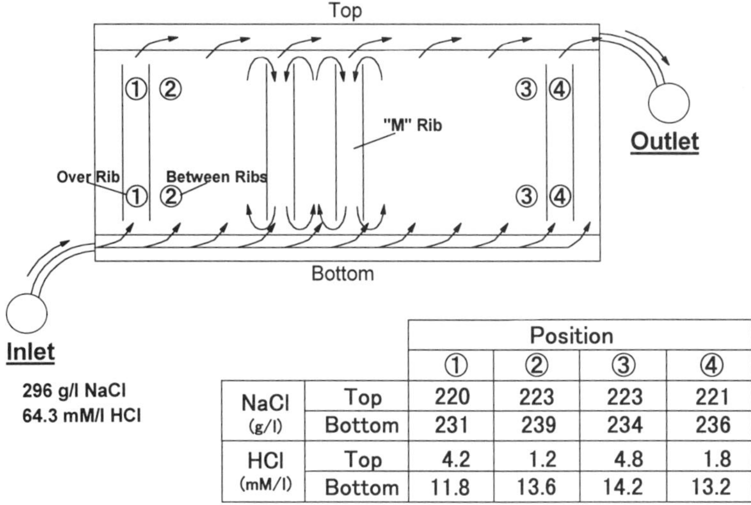

		Position			
		①	②	③	④
NaCl (g/l)	Top	220	223	223	221
	Bottom	231	239	234	236
HCl (mM/l)	Top	4.2	1.2	4.8	1.8
	Bottom	11.8	13.6	14.2	13.2

Figure 14 *Homogeneous Electrolyte Distribution (AZEC-B1)*

3.2.2 Smooth Electrolyte Discharging System. At higher current density, smooth discharge of the gas and electrolyte becomes more difficult due to an increase in volume of the generated gas. A bad discharging system brings bad membrane performance, such as blistering problems due to chlorine gas stagnation, or abrasion problems due to pressure fluctuations. As described in the previous section, in the AZEC-B1 the electrolyte with gas is discharged and disengaged very smoothly by the Dam type overflow. This well designed method attains the object of no gas stagnation and no membrane vibration even at high current density operation of 5 kA/m^2 .

3.2.3 Corrosion Protective Design. Based on long experience of AZEC operation, an anti-corrosive electrode and coating system, etc. are adopted in the AZEC-B1 design. In addition, optimization is made among many points such as maximum cell numbers per electrolyzer, dimension of feed and outlet flexible pipes and piping materials, to avoid current leakage or electrical corrosion. Furthermore, the special welding method used to make the back-plate from the titanium/nickel sheets secures an undiminished and reliable electric connection. The titanium pan and the nickel pan are connected very rigidly with special fusion metal in between. This joint makes the basic structure very reliable for very long term use.

3.2.4 Low Operating Pressure. AZEC-B1 has adopted low operating pressure, the same as used for the AZEC-monopolar type, considering high safety from the mechanical and operational points of view. This results in high reliability for long term operation. At 5 kA/m^2 operation, +450 mmH_2O for the hydrogen gas side and 0 mmH_2O for the chlorine gas side are standard.

3.2.5 Low Power Consumption. In order to realize a stable and long term low power consumption, the electrode gap has been optimized through many tests using pilot cells. Figure 15 shows the expected power consumption based on the actual results obtained from a commercial scale AZEC-B1 electrolyzer. The dotted parts of the line indicate our trial experience for increasing the current density up to 6 kA/m^2. In a full size pilot cell, essential data at 6 kA/m^2 are in the process of being collected, therefore, the performance data for long term operation with a commercial electrolyzer will be reported in the near future.

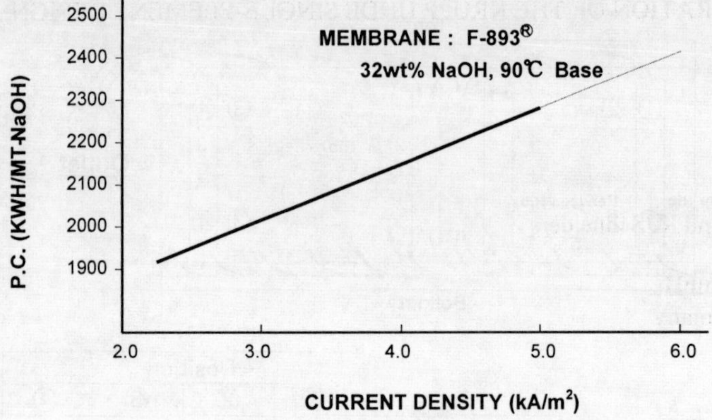

Figure 15 *Power Consumption of AZEC-B1 Electrolyzer*

3.3 Operation of a Commercial Scale AZEC-B1 Electrolyzer

In 1990 a commercial scale AZEC-B1 electrolyzer consisting of 50 elements started successfully at the Asahi Glass Kashima factory. Since then, AGC has entered into demonstrating a long-run operation to prove its excellent performance and durability at high current density. During the first 6 years the cell was operated at 4 kA/m^2 current density. Then, the current density has been increased to 5 kA/m^2 since May 1996. As for the membrane, Flemion F892 was installed for run-#1; but for run-#2 and run-#3, F893 was installed. In spite of the two times membrane replacement, the major portion of the cell elements including anode and cathode have been reused. Figure 16 shows the change of the power consumption for each run. The power consumption increases with a lapse of time, due to both an increase in voltage and a decrease in current efficiency. The voltage increase during the last 7 years is estimated to be approximately 150 mV, and it can be explained by an increase in the cathode over-potential. This increase has mainly occurred at the unscheduled shutdowns during the first couple of years; however, after adopting the use of polarization current during shutdowns, the voltage has been much more stable than before.

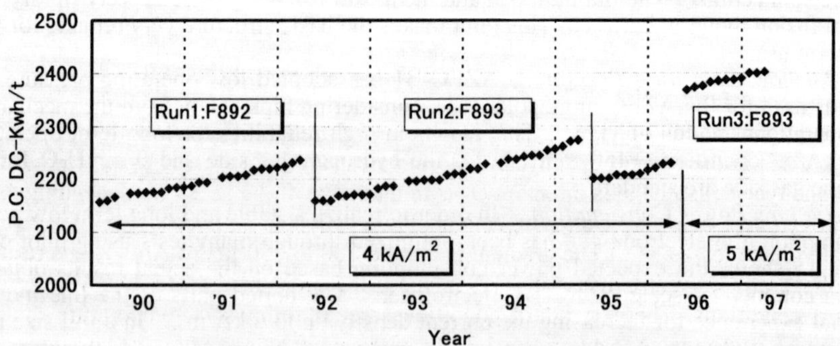

Figure 16 *Performance of AZEC-B1 Electrolyzer in Kashima Factory*

13

A NEW GENERATION OF THE KRUPP UHDE SINGLE-ELEMENT DESIGN

T. Borucinski and K. Schneiders

Krupp Uhde GmbH
Dortmund, Germany

1 PREFACE

Krupp Uhde improved the existing cell design[1] with a view to optimising the membrane electrolyser. The more than 50 plants sold represent a total of over 23000 elements with a total capacity of about 2800 kt_{NaOH}/a. This was the reason why the further development of this design retained the single element principle of the cell jointly developed by Hoechst, Bayer and Krupp Uhde.[2] This cell generation is particularly suitable for operation at current densities of about 5 kA/m^2 and, when required, the oxygen content of the anode gas can be lowered by acidifying the make-up brine. This element type was optimised in order to minimise the concentration differences in the cathode and anode chambers, to minimise the contact and structural losses, to minimise the cell voltage, and to homogenise the current density distribution.

2 INTRODUCTION

The current density/voltage curve of a chlor-alkali electrolysis cell can be expressed with reasonable accuracy for current densities of over 1 to 1.5 kA/m² by the following linear equation:

$$U = U_0 + k \cdot i \tag{1}$$

where U [V] is the cell voltage, U_o [V] is the axis intercept, k [V/(kA/m²)] is the gradient and i [kA/m²] is the current density. In chlorine electrolysis processes using conventional membrane type cells, the axis intercept U_o is in the range of 2.3 to 2.7 V, depending on the technology (steel or nickel cathodes, activated/non-activated cathodes, type of activation, internal voltage losses, etc.).The gradient is around 180 to 250 mV/(kA/m^2). The resulting cell voltage at a current density of 3 kA/m^2 is in the order of 2.84 to 3.45 V referred to standard conditions. These standard conditions are as follows: a temperature of 85 to 90 °C and a catholyte concentration of 32 to 33 % wt. The anolyte concentration (18 to 19 % wt.) is often included in the standard conditions.

As can be seen from equation 2, the specific energy demand $E_{spec.}$ [kWh/t_{NaOH}] is directly proportional to the cell voltage:

$$E_{spec.} = \frac{U \times 100000}{1.492 \times CE} \quad (2)$$

where U [V] is the cell voltage and CE [%] is the current efficiency.

Hence, there is an optimum current density for every cell type, at which the production costs of the cell are lowest. For the Krupp Uhde standard cell,[1] the relative costs per ton of caustic soda solution produced in a 50000 tpa plant were calculated as a function of the current density for different electricity prices and k values of the current density/cell voltage curve. This calculation only took into account the costs of the cell room and the rectifier.

As can be seen from Figure 1, the minimum of the relative production costs as a function of the current density depends on both the electricity price and the k value of the current density/voltage curve. The lower the electricity price and the k value, the higher is the current density at which the production costs are the most favourable. This correlation always applies irrespective of the technology, even if the position of the minimum point and the course of the curve differs for different types of electrolyser. It shows that, irrespective of the electric power costs, high current densities are only economical if the gradient of the current density/voltage curve is small.

3 THE NEW CELL GENERATION

As opposed to the filter press system of other technologies, the Krupp Uhde electrolysis cell consists of single elements in modular construction. Each single element consists of a cathode and an anode half shell, the spaces enclosed are segregated by the membrane. The two half shells are equipped with the required electrolyte inlets and outlets located at

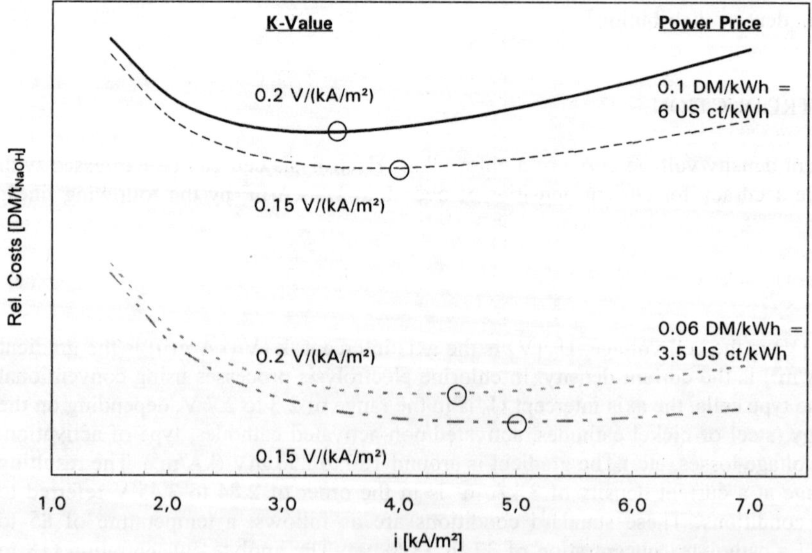

Figure 1 *Relative product costs for a 50000 tpa plant as a function of the current density for different electric power prices. Only the costs of the cell room and rectifier were taken into account. The minimum for each case is marked by a circle.*

opposite ends of the half shells. The edges are sealed by means of a special sealing system and the element bolted together by means of flanges. These elements are suspended in a frame and pressed against each other by a tensioning device in order to achieve the best possible electric contact.

In the further development of the Krupp Uhde cell, the specific feature of modular construction on the single element principle was retained. Emphasis was placed on the following points:

- minimum concentration differences in the cathode and anode chambers
- minimum contact and structural losses
- minimum cell voltage
- homogeneous current density distribution.

4 CHARACTERISTIC FEATURES OF THE NEW CELL GENERATION

Figure 2 shows a vertical section of a single element of the latest type.[2] Solid plates are used to connect the rear wall and the electrode. These plates are joined to the rear wall and the electrode over their entire height.

As opposed to the conventional resistance welding procedure used in the past, the elements are now fabricated using laser beam welding with the aid of high-power CO_2 lasers. The advantage of this welding method is the considerably lower heat input during fabrication, thus minimising the thermal load to which the elements are subjected and reducing the fabrication tolerances. This also has a positive effect on the operation of the elements, as the membrane and all other components are subjected to a lower mechanical load in operation. Moreover, exact electrode spacing is achieved over the entire electrode surface.

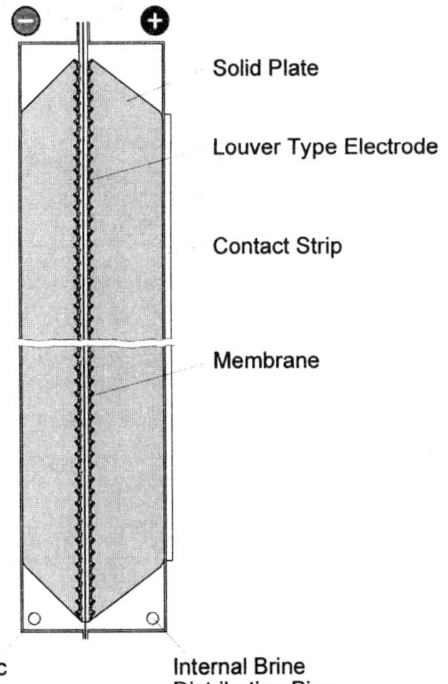

Figure 2 *Vertical section of a single element of the new type.*

The electrolytes are no longer fed to the cell at one single point but through a distributor arranged at the bottom of the element in the half shell. This distributor is used both on the catholyte and on the anolyte side.

The single elements are suspended in a frame for operation and pressed together by a tensioning device in order to achieve an optimum electrical contact with minimum contact voltages between the elements. In a bipolar arrangement, a titanium pan (anode) is always pressed against a nickel pan (cathode). A direct pressure contact between two single elements would result in unacceptable contact losses, because titanium is always coated with a passive non-conducting oxide layer. This is why explosion-plated titanium/nickel contact strips are welded to each anodic rear wall. In this way only nickel/nickel pressure contacts are created between any two single elements, such contacts being characterised by continuous low contact losses. Thanks to the contact strips on the rear wall of the anode half shell, the contact losses in the standard element hitherto used are in the order of a few mV. In order to reduce these contact losses further still, the number of contact strips in the laser-welded element was increased.

By an appropriate adaptation of the welding parameters, the contact strip, the rear wall and the solid plate can be welded together in one operation. This optimised process yields precisely fabricated elements and reduces the fabrication time for the elements. In addition, the integral joint between contact strip, rear wall, solid plate and electrode makes for minimum electrical losses.

As can be seen from Figure 3 below, the symmetrical construction chosen for the new cell generation results in a homogeneous current density distribution.

The upper part of Figure 3 shows the anode half shell of the first element, the cathode half shell at the bottom being part of the second element. The current flows from the cathode via the solid plate to the rear walls and the contact strip and thence into the adjoining anode half shell. From there it flows through the solid plate into the electrode. Because of the symmetrical current entry into the electrode, the current density distribution

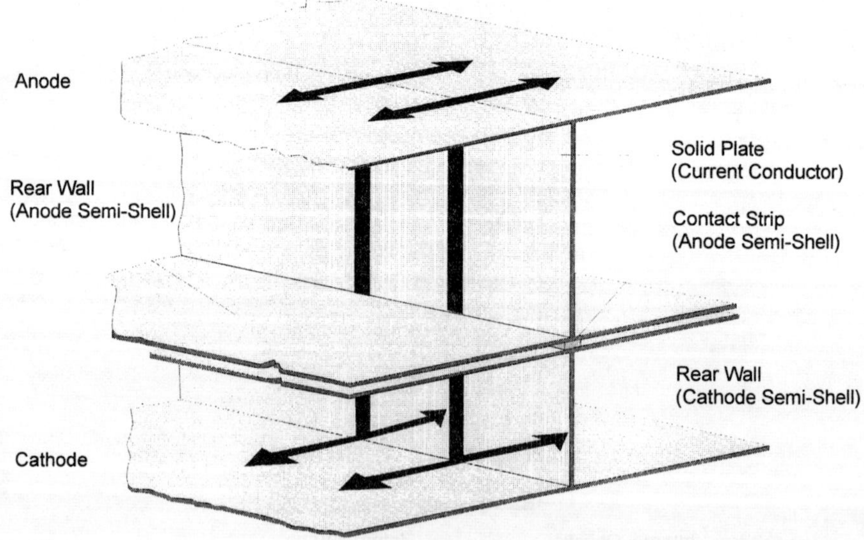

Figure 3 *Three-dimensional section of a single element of the new type. The arrows illustrate the homogeneous current density distribution.*

in this design is completely homogeneous. This means that one of the most important prerequisites for the use of these cells at high current densities has been met.

5 INTERNAL STRUCTURAL AND CONTACT LOSSES

The internal structural losses and the contact losses were determined for a standard element and for an element of the new generation. Figure 4 shows the structural and contact losses of a standard element and a laser-welded element.

Figure 4 shows that the overall losses are reduced by approximately 60 %, which means that these now amount to only around 23 mV at a current density of 3 kA/m^2. At 5 kA/m^2, this represents a lowering of the specific energy demand by about 40 kWh/t_{NaOH}.

6 ELECTROLYTE INLET AND CONCENTRATION DISTRIBUTION

The electrolytes enter the standard element on both the anode and the cathode side at one single point of the half shell. In the case of the laser-welded single element, both half shells are equipped with an electrolyte distributor. This distribution system is necessary, because the half shell is essentially divided into chambers by the solid plates located between the electrode and the rear wall. Moreover, this is the only way to minimise the concentration differences within the cell further still and to approach the conditions of an ideal agitated vessel.

The following illustrations show the concentration distribution measured in operation for a standard element (Figure 5) and for a new element type (Figure 6). The measurement is made in both cases at an anolyte depletion of 80 g/l.

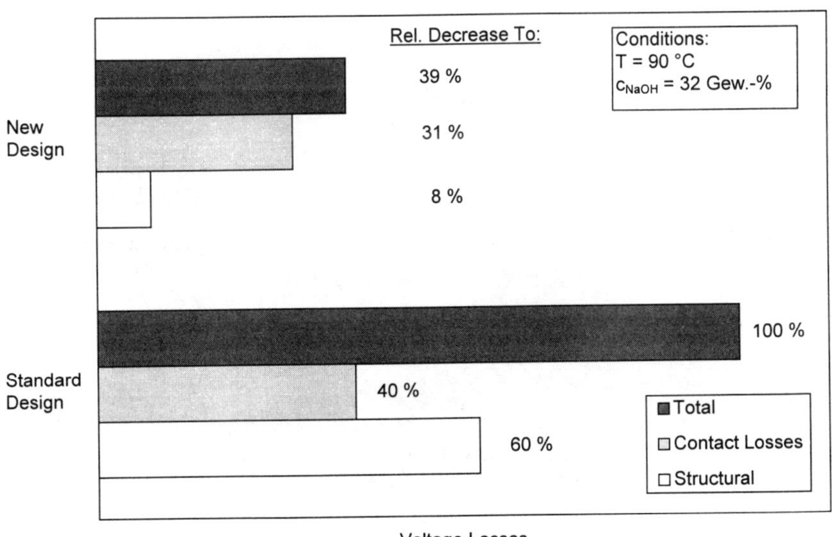

Figure 4 *Structural and contact losses for a standard element and a laser-welded element. The standard type is used as the reference, i.e. the sum of structural and contact losses is 100%.*

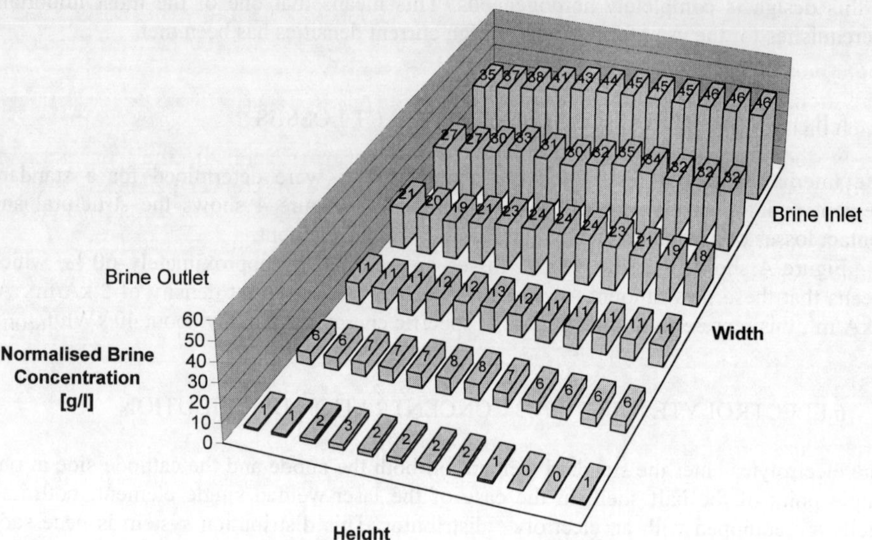

Figure 5 *Concentration distribution in the anode compartment of the Krupp Uhde standard element* ($\Delta c_{Anolyte} \approx 80$ *g/l*).

The illustrations show the difference between the concentration found at a particular point and the minimum concentration encountered in the half shell. The base areas of the illustrations correspond to the cell area; the Z axis shows the normalised concentration differences. In Figure 6, the inlet distributor is represented by a line.

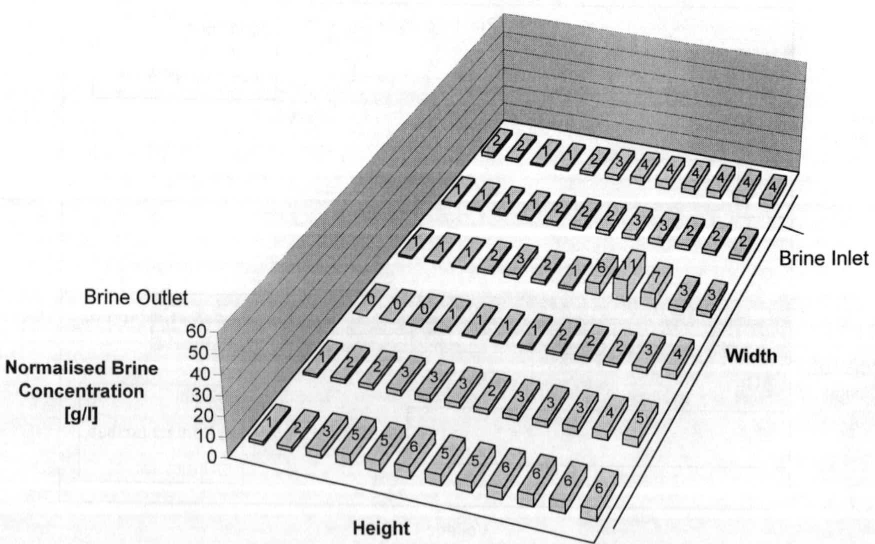

Figure 6 *Concentration distribution in the anode compartment of the laser-welded Krupp Uhde cell* ($\Delta c_{Anolyte} \approx 80$ *g/l*).

In the standard element (Figure 5), the concentration within a row is almost constant, this being attributable to the downcomer effect.[3/4/5] The concentration decreases from the single point of brine entry to the outlet. The maximum concentration differences found were around 45 g/l. This rather low value corresponds to about 60% of the depletion at which this test was carried out.

In the case of the laser-welded single element (Figure 6), the concentration distribution is homogeneous, the maximum concentration differences being less than 10 g/l. This distribution can only be achieved by the selected combination of the inlet system with the downcomer and almost corresponds to the distribution expected in an ideal agitated vessel.

7 CURRENT DENSITY/VOLTAGE CHARACTERISTIC

The new element generation is operated in a high-performance test bay at a continuous current density of 5 kA/m^2. For the purpose of comparing the electrical behaviour of these elements with the standard element used hitherto, the current density/voltage characteristic was determined for the two types in a direct comparison. The curves shown in Figure 7 are standardized and refer to standard conditions of 90 °C and 32 %-wt. NaOH.

The illustration shows that the linear relationship for voltage and current density described earlier does not depend on the element type. Using the same membrane, the standardized voltage at a current density of 5 kA/m^2 differs by about 120 mV.

By using suitable high-efficiency membranes, the cell voltage can be lowered further still. Compared with the standard design, a total of about 170 mV can be saved at a current density of 5 kA/m^2, which represents a lowering of the specific energy demand by 120 kWh/t_{NaOH}.

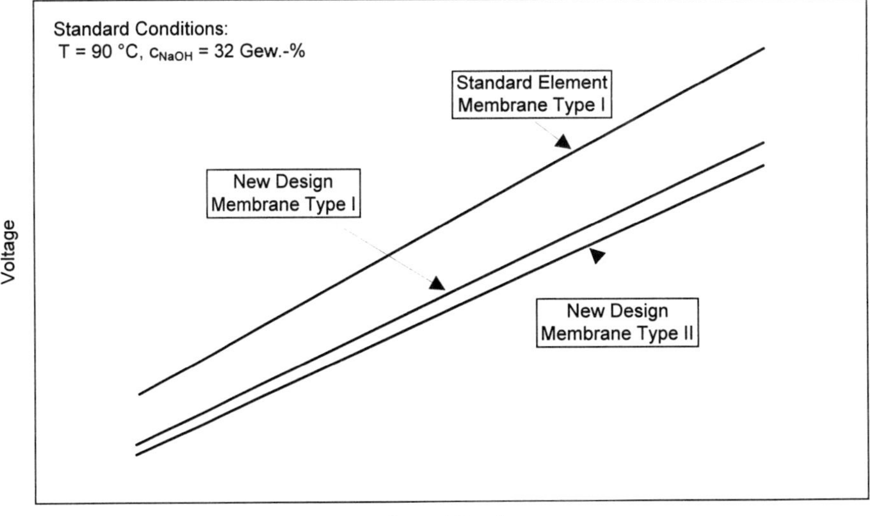

Figure 7 *Current/voltage characteristic for different element and membrane types (90 °C, 32 % wt. NaOH).*

8 DISCUSSION

At a current density of 5 kA/m² and using the same membrane type, approximately 120 mV can be saved with the laser-welded cell. Measurements of the structural and contact losses show that about half this saving can be attributed to a lowering of the voltage losses within the element. The remaining 60 mV are due to the optimisation of the concentration distribution and the resulting homogenisation of the current density distribution.

The homogeneous concentration distribution with maximum concentration differences of less than 10 g/l can only be achieved by the selected combination of internals (downcomers) and the inlet system. Compared with the external means of circulating the electrolyte by external pumps, the internal circulation induced by the internals has the advantage that the internally induced circulation rates are greater by up to two orders of magnitude and that the resulting homogenisation is more effective. Moreover, no additional control or safety facilities are required.

The symmetry of the cell results in the current being distributed uniformly over the entire electrode surface. This and the minimal concentration differences that prevail create optimum prerequisites for a long membrane service life. In addition, the product quality is improved, because the oxygen content is lowered.

References

1. US Patent: 4,664,770, 1987.
2. German Patent: 196 41 125. 4-45, 1996.
3. German Patent: DE 42 24 492 C1, 1993.
4. European Patent: 93 107 034.6-2111, 1995.
5. R. W. Curry, Modern Chlor-Alkali Technology, Society of Chemical Industry, London, 1995, Vol. 6, p. 197.

14
SIMULATION OF ION BEHAVIOUR IN ION EXCHANGE MEMBRANES

Heiichiro Obanawa, Hideaki Noaki, Hisashi Takei and Hiroyoshi Hoda

Asahi Chemical Industry Co., Ltd.
3-2,Yako 1-Choume,
Kawasaki-Ku, Kawasaki-City,
Kanagawa 210, JAPAN

1 INTRODUCTION

A clear understanding and systematic description of the behaviour of ions in ion exchange membranes is highly desirable, as it would provide a tool for improvement of their design and performance. A great deal of empirical knowledge of the behaviour of electrolytic impurities has been gained through operational experience and laboratory experiment, but a systematic description of their intramembrane behaviour has proved elusive.

Those studies which have been performed on intramembrane ion behaviour have generally been limited to single-layer membranes[1], rather than the two-layer membranes which are used for salt electrolysis, because of the difficulty of describing the behaviour at the interface between their carboxylic acid (C) and sulfonic acid (S) layers.

Here we describe the application of a simulation developed specifically for the two-layer membranes and a comparison of its predictions with experimental findings. The simulation is based on an adaptation of the Nernst-Planck equation in which the conditions at the interface are inferred from measurements obtained in a three-compartment experimental cell.

2 PREDICTION OF INTRAMEMBRANE CONCENTRATION AND pH PROFILES

The Nernst-Planck equation was adopted as an expression of the migration of physical species in ion exchange membranes, which requires the assumption of infinitely dilute conditions but utilizes experimentally measurable coefficients, and a method of simulating the ion behaviour and pH profile was developed on this basis.

The migration of physical species in the ion exchange membrane can be expressed in terms of the Nernst-Planck equation, as shown Table 1. It represents the driving forces for the migration of the species through the membrane, in terms of (1) electrical transport based on the electric potential gradient, (2) diffusion based on the concentration gradient, and (3) water transport.

In this equation, the concentration profiles in the membrane can be calculated by finite differential methods, since the membrane interior, including the fixed-ion groups, can be taken as electrically neutral, and the water-dissociation equation can be utilized.

Studies on intramembrane ion behaviour have generally been limited to single-layer membranes, rather than the two-layer membranes which are used for salt electrolysis, because of the difficulties of describing the behaviour at the interface between carboxylic acid and sulfonic acid layers.

The interface between the carboxylic acid and sulfonic acid layers of the membrane is a region of sharp discontinuity in terms of mathematical treatment. However, the pH and ion concentration profiles in this region are essential for simulation of the overall profile. To overcome this problem, a special three-compartment laboratory cell was developed (Figure 1). The middle compartment is bordered on one side by a

Table 1 *Nernst-Planck equation*

$$Ni = -Zi \cdot \frac{F \cdot Di}{R \cdot T} \cdot Ci \cdot \nabla\phi - Di \cdot \nabla Ci + Ci \cdot V \quad \text{(eqn 1)}$$

$$Zm \cdot Cm + \Sigma\, Zi \cdot Ci = 0 \quad \text{(eqn 2)}$$

$$C_{OH^-} \cdot C_{H^+} = Kw \quad \text{(eqn 3)}$$

Where:

Ni : Flux Zi : Charge number Ci : Concentration

ϕ : Electric Potential V : Velocity R : Gas constant

Di : Diffusion coefficient F : Faraday's constant

T : Temperature Kw : Water equilibrium constant

Suffix i : Species m : Fixed ion

single-layer sulfonic acid membrane, and on the other by a single-layer carboxylic acid membrane. The pH and concentrations in all three compartments were measured, and the values at the membrane surfaces in the middle compartment at equilibrium were assumed, and were taken as the values applying at the interface in the actual two-layer membrane, in the overall simulation.

Figure 2 shows the results obtained with the three-compartment cell, under various operating conditions. As shown by Figure 2a, increasing the concentration of NaCl in the anode compartment resulted in a proportional increase in the concentration of Cl^- in the middle compartment. It shows that diffusion is the governing mechanism of Cl^- movement through the sulfonic acid layer. As shown by Figure 2b, the minimum OH^- concentrations in the middle compartment were observed with cathode compartment concentrations around 33% caustic.

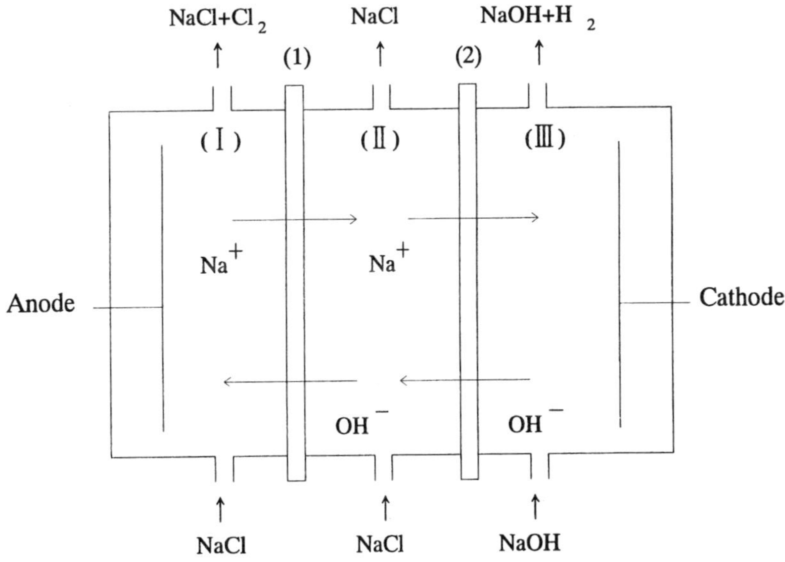

Figure 1 *Three-compartment cell*

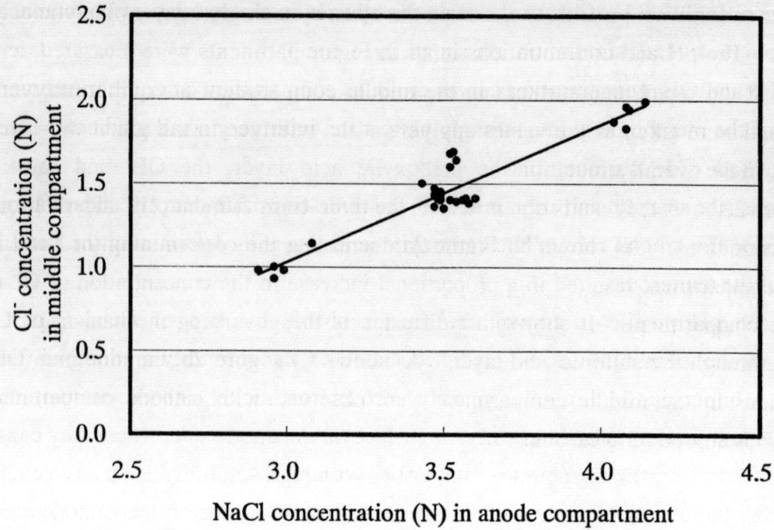

Figure 2a *Influence of brine concentration on Cl^- concentration in middle compartment*

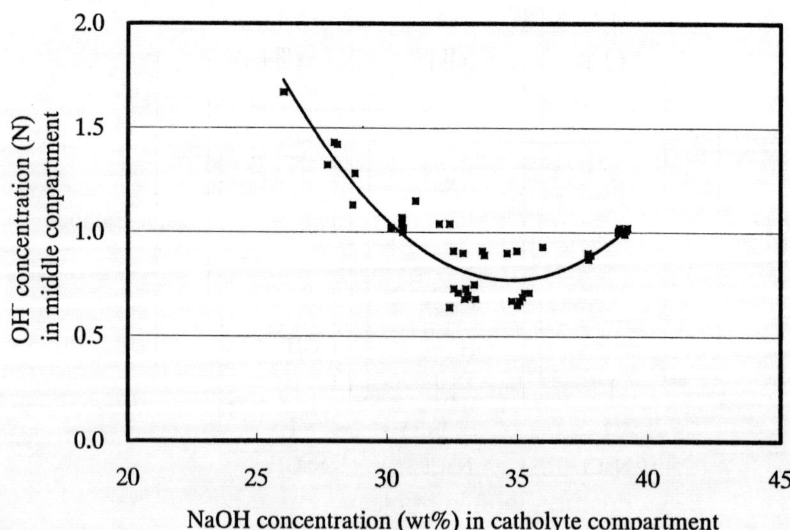

Figure 2b *Influence of caustic soda concentration on OH^- concentration in middle compartment*

The concentration profiles of the main ion species in the two-layer membrane, as obtained from the overall simulation, are shown in Figure 3.

The H^+ and OH^- concentration curves cross, as soon as they enter the membrane. Almost all of the membrane is in a strongly basic state, with very small gradients. But near the cathode-side surface, in the carboxylic acid layer, the OH^- and caustic concentrations rise sharply, and the Cl^- and H^+ concentrations fall sharply. These large gradients show the crucial role of the carboxylic acid layer, in determining the current efficiency of the membrane.

Once the pH profile is known, the most probable location of metal impurity depositions in the membrane can be predicted, based on the solubility products of their hydroxides. Impurities having low solubility, such as iron, nickel, and magnesium, tend to precipitate and deposit near the membrane surface on the anode side, where they cause a rise in the IR drop of the membrane. Impurities with high solubility, such as calcium, strontium, and barium, tend to deposit near the membrane surface on the cathode side. This leads to the formation of large voids, which cause declines in current efficiency. The actual situation is more complex, of course, since various impurities tend to interact and form complex compounds.

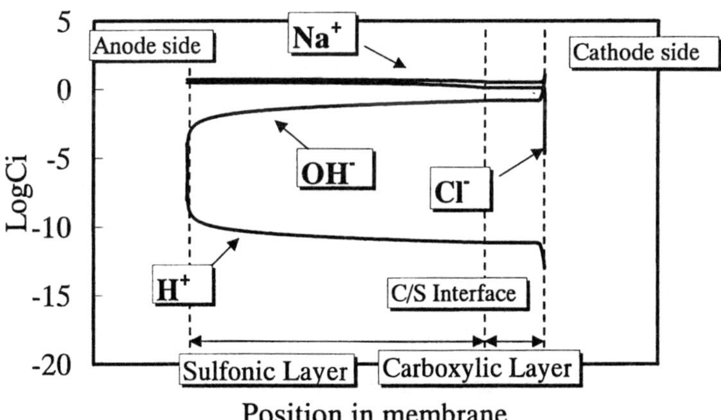

Figure 3 *Intramembrane ion concentration profile (calculated)*

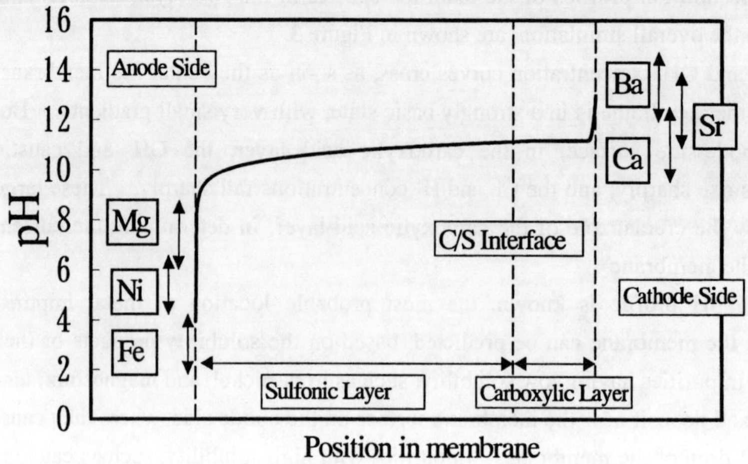

Figure 4 *Map of brine impurity deposition (calculated)*

3 PREDICTION OF IMPURITY DEPOSITS

3.1 Behaviour of Fe^{3+}

For convenience, Fe^{3+} was chosen for the initial comparison of the predicted and experimental behaviour of ion impurities in the membrane, because of its low solubility product and its relatively small influence on current efficiency and membrane voltage.

The occurrence and location of $Fe(OH)_3$ deposits in the membrane were predicted on the basis of the $Fe(OH)_3$ solubility product (3.98×10^{-41} at 100°C, indicating its precipitation at pH 2.4) and the pH profiles obtained in the Nernst-Planck simulation, for brine containing 100 ppb Fe^{3+} at various acidities and operation at 90°C, 4 kA/m^2, and 23 wt% caustic soda. The simulation therefore predicts that no significant deposits will normally occur in the absence of HCl addition. With anolyte containing 0.008 N HCl, deposits will be located at about 2 μm from the membrane surface. With anolyte containing 0.025 N HCl, the region of deposits will be located at about 8 μm from the membrane surface (Figure 5).

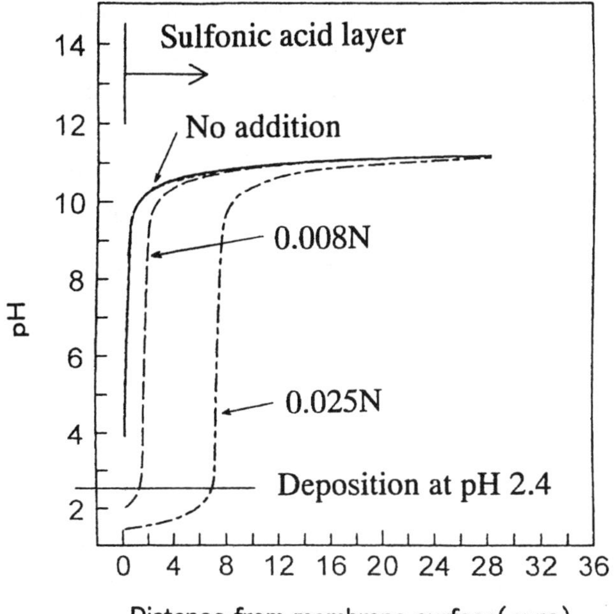

Figure 5 *Effect of HCl addition on intramembrane pH profile (calculated)*

(NaOH 23% / NaCl 2.75 N / 4 kA/m² 90 °C)

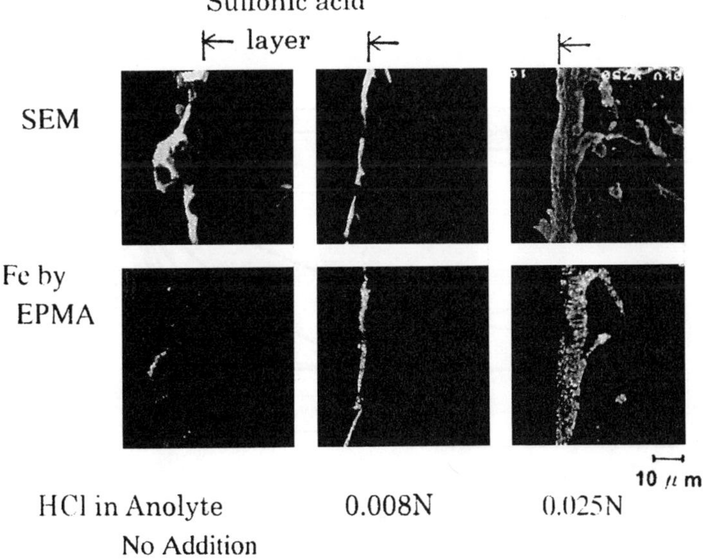

Figure 6 *Location of Fe in membrane cross-section (NaOH 23% / NaCl 2.75 N / 4 kA/m² 90 °C)*

These predictions were then compared with the Fe^{3+} deposits actually observed by electron probe microanalysis (EPMA) following the operation of an experimental cell under the same conditions for 90 days (Figure 6). With brine containing no added HCl, precipitation occurred largely within the anolyte compartment and in slight amounts on the face of the membrane facing it. With the addition of 0.008 and 0.025 N HCl, however, Fe^{3+} deposits were found at 2 μm and 10 μm from the membrane surface. The predictions obtained from the simulation were thus in close accord with the observed results.[2]

3.2 Behaviour of Ca^{2+}

It is well known that electrolysis with brine containing Ca^{2+} tends to result in its deposit in the membrane near the cathode and a consequent lowering of current efficiency.

In the simulation using the Nernst-Planck equation, the slope of the pH profile near the surface of the membrane facing the cathode increases with increasing current density, and the onset of this rise in pH thus approaches the membrane surface (Figure 7). Accordingly, the simulation predicts that increasing the current density should result in an increasing proximity of Ca^{2+} deposits to the membrane surface.

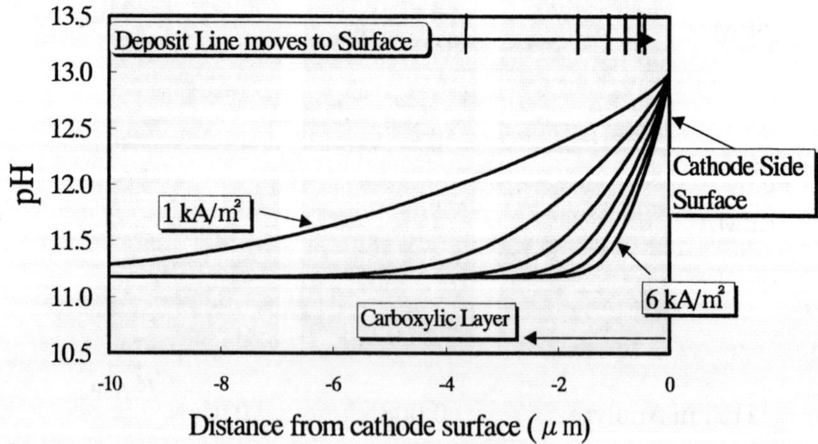

Figure 7 *Effect of current density on intramembrane pH profile (calculated)*

These predicted results were also in accord with experimental observation (Figure 8). Operation was conducted with brine containing 3 ppm Ca^{2+}, under operating conditions of 90 °C, 3.5 mol/dm^3 brine, 33 wt% caustic soda, and 1, 2 or 4 kA/m^2 current density. The position of the Ca^{2+} deposits in the membrane were then measured by EPMA.[3]

4 CONCLUSION

The results of the present study indicate that simulation based on the Nernst-Planck equation to obtain the intramembrane concentration and pH profiles, together with known solubility products, provides an effective means of predicting the behaviour of ions and deposit of impurities in the two-layer salt electrolysis membrane, and a valid basis for explaining this behaviour, predicting impurity deposits and their effect on membrane performance, and improving membrane design and tolerance to impurities.

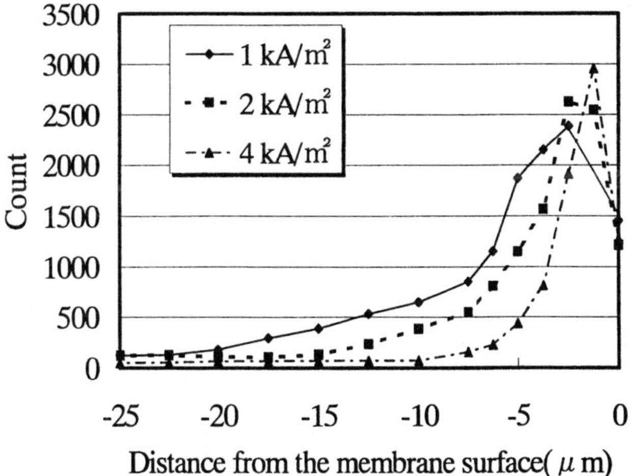

Figure 8 *Calcium deposits in membrane following experimental operation, as determined by EPMA*

The predictions are also in accord with experimental measurements, and with observations in a broad range of commercial operations for caustic soda production.

Further study is necessary. Actual operating conditions are more complex than those described here, as they involve the interaction of various impurities. We will continue to work for a greater understanding and description of these complex systems, as part of the continuing effort for new advances in ion exchange membranes for salt electrolysis.

References

1. G. Pillay, and D.L. Galdwell, R.E. White, The 183rd Meeting of E.C.S, The Electrochemical Society, Inc., Pennington, N.J., USA, 1993, p.1856.
2. Y. Noaki, H. Takei, H. Hoda and Y. Shiroki, Proceedings of the 18th Symposium on Caustic Soda Technology (Japan), The Electrochemical Society of Japan, Tokyo Japan, November 1994, p.90.
3. H. Takei, and H. Hoda, Y. Noaki, M. Hamada, Proceedings of the 19th Symposium on Caustic Soda Technology (Japan), The Electrochemical Society of Japan, Tokyo Japan, November 1995, p.66.

15

THE FORMATION OF PRECIPITATES OF IRON IONS INSIDE PERFLUORINATED MEMBRANES DURING CHLOR-ALKALI ELECTROLYSIS

Ir. J. H. G. van der Stegen,

Akzo Nobel Central Research B.V. Arnhem

Ir. P. Breuning,

Akzo Nobel BU SALT. Amersfoort

1 INTRODUCTION

The tolerance for iron of perfluorinated membranes in chlor alkali electrolysis is a matter of controversies. In 1979 Akzo Nobel acquired a license for a plant equipped with membranes from Asahi Chemical, for which it had to meet a specification of less than 20 ppb of iron in feed brine to the cells. Later on (1985) Dupont made the following statement: "Iron hydroxide forms in anolyte, collects on anode and membrane surfaces". It is thought that iron can be tolerated, due to the insolubility of iron hydroxide up to 1 mg/l.[1-3] More recently, Naoki *et al.*[4] discussed the influence of anolyte acidity and showed iron penetrating into the membranes, depending on acidity. The present paper therefore has for its object to clarify the controversies. More particularly that it is crucial whether the membrane is operated in such mode that the pH at its anolyte surface is in between 5 and 11 resulting in an insoluble iron hydroxide.

2 SOME HISTORY IN BRIEF

In 1979 Akzo Nobel acquired a license from Asahi Chemical on a plant whose present annual production amounts 300,000 t chlorine. The plant was built in Rotterdam and started operation in 1983. The plant is operated using forced anolyte circulation and anolyte acidification. To meet the specified performance it was required by the licensor Asahi Chemical in 1979 that the feed brine should contain less than 20 ppb of iron.

Today's performance of Aciplex membranes in the plant was reported earlier by Shiroki *et al.*[5] The cell voltage does not increase over 50 months of membrane operation. Over the same period the current efficiency decline is 2%. These results are among the best in the world.[6] So there is no doubt about plant performance as long as feed brine contains less than 20 ppb of iron.

In this paper it will be shown, by experimental results in which ferric chloride was added to the feed brine that iron may cause under circumstances negative effects on cell voltage or current efficiency of chlor alkali membrane electrolysis.

Akzo Nobel studied the effects in plants where a change of feedstock and brine treatment caused the iron, contained in feed brine, to rise from < 5 ppb to on average 200 ppb. This increase of the iron content was considered to be safe as membrane suppliers,[1-3] since 1985, allow the feed brine to contain up to 1 mg/l of iron. Membrane suppliers based their opinion on the fact that iron hydroxide is insoluble and accumulates at the membrane surface. Some laboratory tests support this view.

3 OBSERVATIONS IN INDUSTRIAL CHLOR ALKALI PLANTS

The experimental evidence, on which the present paper is based, was obtained in several industrial tests. The change in iron content in feed brine resulted from a change in salt feed. More in particular the plants changed from solar salt or rock salt to evaporated salt. Table 1 presents the typical composition of the salt.

The plants, in which the tests were run are equipped with brine treatment systems capable of treating the calcium, magnesium and sulfate. Sulfate is precipitated, either as calcium sulfate or barium sulfate. Calcium and magnesium are precipitated as calcium carbonate or magnesium hydroxide.

The available equipment consists of:

- reactors in which precipitates are formed,
- dosing equipment for adding the chemicals needed ($NaOH$, Na_2CO_3, $BaCO_3$ or $CaCl_2$),
- a settler to remove precipitates,
- an alfa cellulose - precoated filter to remove fines processed at pH = 10,
- an ion exchanger unit to catch the remaining soluble calcium and magnesium ions to below the 10 ppb level in feed brine.

When changing the feed stock from rock salt or solar salt to evaporated salt the brine purification can be bypassed completely. There is no longer need for addition of the chemicals. Impurities like iodine, barium, silicates which are known to be possible causes of membrane failure are no longer available from the salt feed. In all cases of all kinds of salt feed calcium, magnesium and ferric levels in feed brine are controlled by filtration and by the ion exchanger unit.

However when the anticaking agent, potassium ferrocyanide, is present in the salt and is not decomposed then complex iron passes through the filters and ion exchangers. It will be decomposed in the electrolysers; and the membrane is exposed to the effect of its decomposition product: iron ions. The majority of these iron ions leave the electrolyser with depleted brine and the precoated filter will remove these iron ions after resaturation. Table 2 presents the typical feed brine composition during the tests.

In plant A in which the changeover was made from rock salt to evaporated salt typical cell voltage at standard conditions rose from 3.2 volts to 3.6 volts so 400 mV in 18 months. In plant B the changeover was made from solar salt to evaporated salt and there the test lasted three months and the cell voltage rose by 60 mV maximum.

Table 1 *Composition of salt feed stock during the tests*

element	*rock salt*	*solar salt*	*evaporated salt*
sulfate in ppm	19000-21000	1500-1800	350
calcium in ppm	9000-10000	400-1500	5-12
magnesium in ppm	500-700	50-1000	<1
ferrocyanide in ppm	none	none	7
insolubles in%	2-4	<0.1	<0.01

Table 2 *Typical feed brine compositions*

impurity	*rock salt*	*solar salt*	*evaporated salt*
sulfate g/l	3-6	3-6	3-6
calcium ppb	10-15	10-15	10-15
magnesium ppb	<1	<1	<1
iron free ppb	<10	<10	<10
iron complex ppb	<1	<1	200
iron total ppb	<10	<10	200
other	same	same	same

During the period of 200 ppb iron in feed brine after each power shut down, both plants observed a drop in voltage as if something had fallen off the membrane or anode surface.

Both plants are operated without acidification of the anolyte. The cell designs of both plants were such that feed brine of 300 g/l strength entered the anolyte compartment. Depleted brine left the anolyte compartment containing 200 g/l NaCl. The cell designs in both plants were such that internal mixing of the anolyte in the anolyte chambers is poor.

In a third plant a similar change was made with ferrocyanide decomposition and no cell voltage effects were observed.

The two plants in which the tests were run without ferrocyanide decomposition changed back to the original salt stock and the voltage effects disappeared. These results are depicted in Figures 1 and 2. They induced Akzo Nobel to carry out a more extensive investigation.

The extended investigation covered ten industrial plants and an industrial test facility of four different electrolysers. These four test electrolysers are operated on one brine quality. These investigations revealed:

(a) Excellent membrane behavior, current efficiency decline $< 0.5\%$ per year and voltage increase < 25 mV per year, was observed in all those cases where feed brine contained less than 20 ppb of iron and no other detrimental impurities.

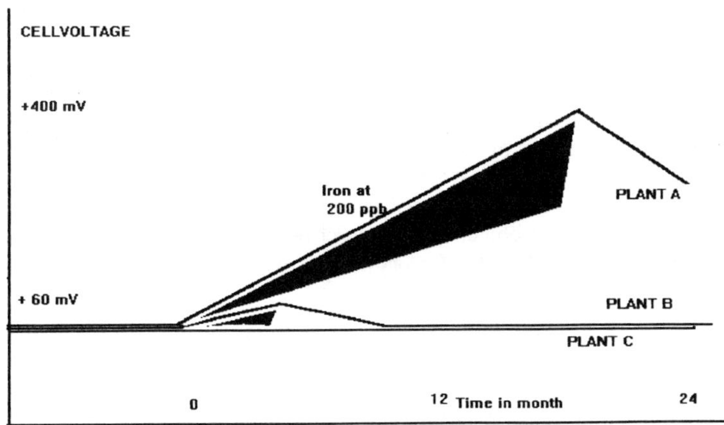

Figure 1 *Voltage increase during industrial tests*

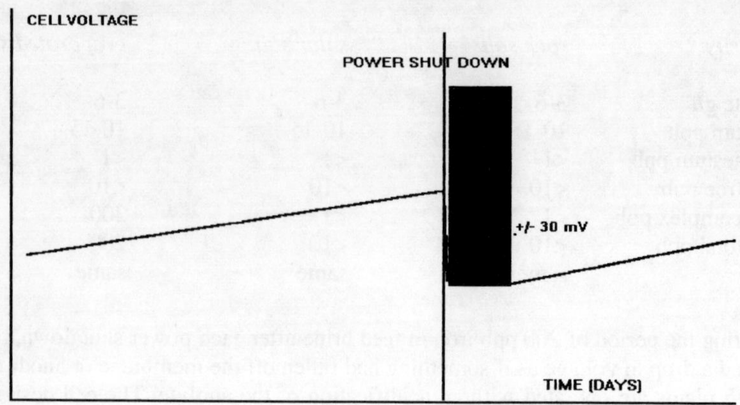

Figure 2 *Voltage drop due to power shut down*

(b) Amongst all impurities mentioned in the feed brine specifications by membrane suppliers,[1-3] only iron can be detected in/on the membrane after use when processing evaporated salt containing ferrocyanide. Iron accumulation inside the membrane was shown with the aid of electron microscopy. In all cases iron hydroxide deposits on the membrane surface are observed.

(c) In some cases the removal of iron from the above mentioned contaminated membranes with the aid of acid resulted in the current efficiency increasing by 3 to 4%. In these cases the polymer seems to be relatively undamaged. The current efficiency decline may be attributed to iron.

(d) There are several membrane suppliers and several membrane types. In some plants several of these membrane types are operated, on purpose of comparison, on one brine quality. This comparison of results will create a ranking of membrane types according to cell voltage increases or current efficiency declines. Some membrane types are found to be sensitive regarding the effect of iron others show a high degree of insensitivity.

(e) The use of different electrolysers on one brine quality, independent whether this brine contains 200 ppb iron or < 20 ppb iron, will result in a ranking of electrolyser designs. In general there is, today, a trend towards membrane electrolyser designs of the bipolar type with strong internal mixing of anolyte and catholyte.

From all these facts it was concluded that the behavior of iron should be described in more detail. Especially, the objective is to describe where and when it precipitates during membrane electrolysis and by what factors the precipitation is governed.

4 SOURCES OF IRON IONS AND THEIR PRECIPITATION INSIDE AND OUTSIDE THE MEMBRANE

Typical conditions in the anolyte chamber of a chlor alkali membrane electrolyser are:

- temperature >90 degree C
- active chlorine 1 g/l
- pH 2 to 4

Under these circumstances ferrocyanide from the salt that passed through the filters and ion exchangers is oxidized to ferricyanide. It is further decomposed and after half an hour only ferric ions, carbon dioxide, and nitrogen or dinitrogen oxide are left.

These ferric ions add up to other sources of iron in the anolyte circulation. The membrane will spot these ferric ions, resulting from both sources.

Two effects occur:

1: Ferric ions enter the membrane, and to which the same rules apply that govern other multivalent cations such as calcium, magnesium, nickel or aluminium. These cations are well known to damage the membrane performance, which is why they are kept below the 20 ppb level specified by membrane suppliers. These cations are known to precipitate as hydroxides inside the membrane according to their solubility products. The effects of these iron ions are described by Naoki *et al.*[4]

2: Ferric ions precipitate on the membrane surface. These deposits of iron hydroxide are observed on used membranes after some months of exposure to anolyte containing iron ions.

As it is the intention of this paper to describe the behavior of iron in more detail, the effect of iron was tested in laboratory cells.

5 EFFECT OF FERRIC IONS UPON THE MEMBRANE IN CHLOR ALKALI ELECTROLYSIS

The following items were a starting point to an experiment in which 200 ppb of iron as ferric chloride was added to the feed brine:

(a) the two general considerations mentioned in the previous paragraph,

(b) the paper of Naoki *et al.*[4] showing the entrance of iron ions into the membrane, depending on anolyte acidity,

(c) the paper by Ogata *et al.*[7] showing the influences of the conditions of operation on the anolyte/membrane interface pH.

As said before, 200 ppb of iron was added to the feed brine as ferric chloride. The test lasted 120 days. During this period of testing the operating conditions (current density, anolyte strength, catholyte strength, acidity of anolyte) were maintained, over periods of ten days, at the set values according to a previously designed experimental program.

Following was observed: at 200 ppb of iron in feed brine the cell voltage rose by 70 mV in 120 days. (Conditions of voltage measurement: 4 kA/m^2; 23% caustic; 180 g/l NaCl; brine acidity 0.02 N H+: membrane with a typical current efficiency of 96%.)

It should be recalled that at 10 ppb of iron in feed brine there is no increase in cell voltage, neither in laboratory cells nor in industrial cells.

The experiment was designed to study also the effect of operating conditions according to a previously designed program. Typical ten day periods were in the test design. The result of the typical effects of anolyte strength and caustic strength on the increase of cell voltage is presented in Figure 3A and Figure 3B. Typical effects of anolyte acidity on both current efficiency and cell voltage are tabulated in Table 3.

It is well known[2,3] that current efficiency increases with anolyte strength. This increase is observed both in the presence or in the absence of iron ions. This effect was also observed, with iron ions present, during the tests. As mentioned before the voltage effect is absent in the absence of iron ions.

It is well known[2,3] that current efficiency reaches with caustic strength a maximum value, thus both with iron ions present and in the absence of iron ions. This effect was observed indeed during the tests.

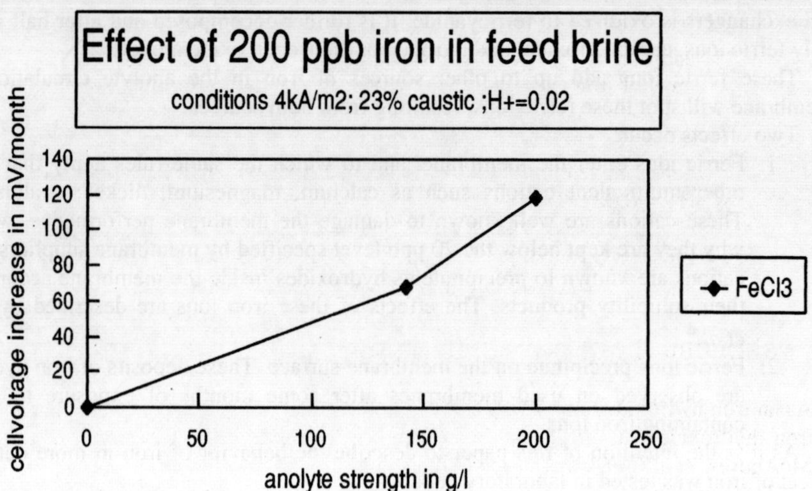

Figure 3A *Increase in cell voltage due to the addition of 200 ppb of iron as ferric chloride. The effect of anolyte strength.*

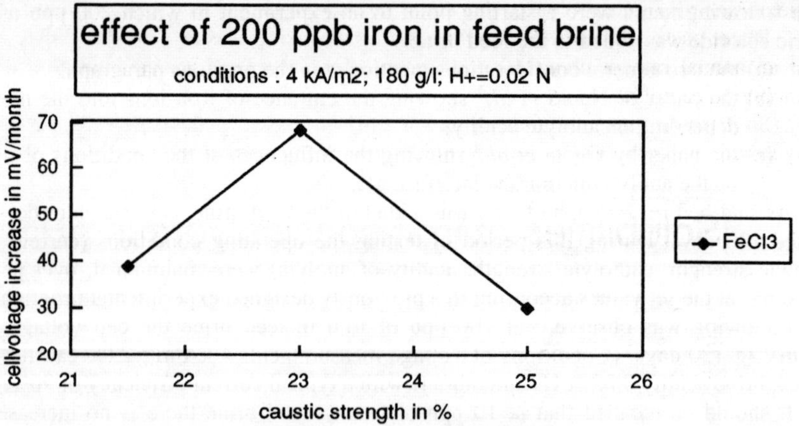

Figure 3B *Increase in cell voltage due to the addition of 200 ppb of iron as ferric chloride. The effect of caustic strength.*

Table 3 presents the results that were obtained during the tests. Changing the acidity of anolyte affects both current efficiency and cell voltage.

These influences of operating conditions complicate the interpretation of effects and might well account for the controversies regarding the maximum allowed level of iron in anolyte. So it will be worthwhile to summarize all possible effects on the role of iron.

The overall test over 120 days showed an increase in cell voltage. The, more detailed, ten day tests showed the influences of operating conditions.

Table 3 *Observed effects of anolyte acidity on trends of cell voltage and current efficiency*

addition of level of iron	ferric chloride 200 ppb current efficiency	ferric chloride 200 ppb cell voltage
non acidified anolyte	- 0.15% per month	nil mV per month
H+ = 0.003 N	nil	+30
H+ = 0.02 N	nil	+68
H+ = 0.06 N	-0.47	nil

6 SOLUBILITY OF IRON HYDROXIDE

As an iron hydroxide deposit is observed at the membrane-anolyte surface, the solubility of iron hydroxide has to be a key issue in understanding the observed phenomena. McAndrew *et al.*[8] reported basics of the solubilities of iron species.

The horizontal axis of Figure 4 represents the pH; and the vertical axis the logarithm of the solubility in mol/l. Figure 4 shows the solubilities of several iron hydroxide species such as: Fe^{+++} ions; $Fe(OH)^{++}$; $Fe(OH)^{+}$; uncharged $Fe(OH)_3$ and anions like $Fe(OH)_4^-$.

So the major factor controlling the solubility of iron hydroxide and the formation of the iron hydroxide deposit is pH. As deposits are formed at the interface of anolyte and membrane pH of electrolyte at that spot, thus the interface membrane-anolyte, is important.

Water passes through the interface and subsequently through the membrane at 4 kA/m^2 at an annual rate of about 60 - 80 m^3/m^2. This water carries all kinds of cations from anolyte to catholyte.

So if iron is present in anolyte and at a pH of above 5 and below 11 at the surface, iron hydroxide will precipitate from the electrolyte that is to enter the membrane.

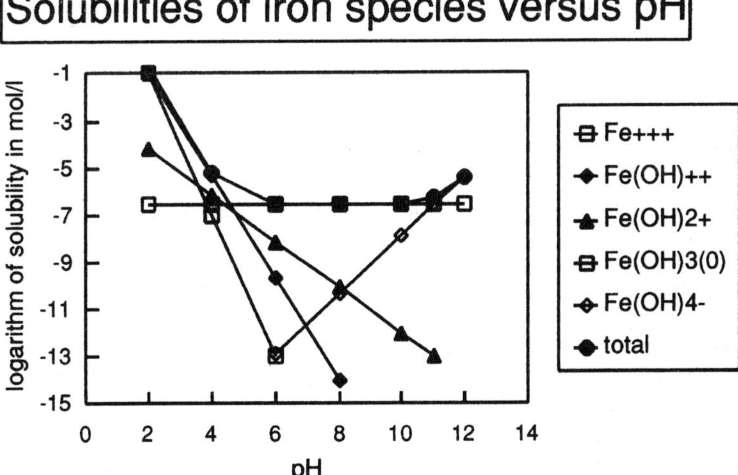

Figure 4 *Solubility of iron hydroxide versus pH*

When the pH is below 5 ferric ions will remain soluble and will be able to enter the membrane along with the electrolyte. Actually, Naoki *et al.*[4] observed penetration into the membrane down to a depth of 10 microns depending on the acidity of the anolyte.

When the pH at the surface rises to above 11, the ferric anions will be available to enter the membrane.

7 pH AT THE SURFACE AND pH GRADIENT ACROSS THE MEMBRANE

During electrolysis the membrane separates anolyte which is weakly acid (pH=2-4) from catholyte, which is strongly alkaline (pH=14). So there must exist a pH gradient across the membrane.

Water passes through the membrane at 4 kA/m^2 at an annual rate of about 60 - 80 m^3/m^2. This water carries the cations from anolyte to catholyte. Precipitation of the hydroxides of these cations is governed by their solubilities and the pH gradient. For reference Ogata *et al.*[9] and Momose *et al.*[10] studied the effects of calcium and magnesium from feed brine. They actually showed their precipitation, depending on their solubilities in relation to pH, somewhere in the interior of the membrane.

Naoki *et al.*[4] applied the same principle to iron. Precipitation in the interior of the membrane or on the surface was shown depending on the acidity of the anolyte used.

The solubility products of the hydroxides of these three cations are:

$[Fe] * [OH]^3 = 10\text{-}35\ (mol/l)^4$ minimum solubility at pH = 7

$[Mg] * [OH]^2 = 4.1 * 10\text{-}12\ (mol/l)^3$ minimum solubility at pH = 11

$[Ca] * [OH]^2 = 9.3 * 10\text{-}7\ (mol/l)^3$ minimum solubility at pH = 12

Iron hydroxide is the first hydroxide of these three to reach its minimum solubility, as the pH increases. So it is the first precipitate that will separate from the electrolyte when it passes through the membrane. Iron hydroxide precipitates are observed in fact on the membrane surface or to a depth of a few microns below it.

Magnesium hydroxide is second on the list of minimum solubility pH and in fact is found in the sulfonic layer of the membrane.

Third is calcium hydroxide, which is in fact observed in the carboxylic layer of the membrane.

8 FACTORS INFLUENCING THE pH AT THE SURFACE AND THE pH GRADIENT INSIDE THE MEMBRANE

Information on the pH at the anolyte membrane interface is presented by Ogata *et al.*[7]

These authors showed:

(a) surface pH values ranging from 7 to 9 under standard conditions,
(b) increasing anodic surface pH with decreasing current efficiency,
(c) decreasing pH with decreasing current density,
(d) increasing pH with catholyte strength,
(e) some effect, either increase or decrease, of the membrane (Flemion versus Nafion) on the anodic surface pH.

These observations are interpreted by comparing:

(a) the availability of protons moving out of the anolyte towards the interior of the membrane by diffusion, migration and convection;
(b) the availability of hydroxyl ions moving out of the catholyte towards the interior of the membrane by diffusion, migration and convection.

The availability of protons depends on anolyte acidity and on the hypochlorous acid concentration. Moreover the influence of anolyte strength upon the activity coefficients of protons should be recognized. In case of a thick iron hydroxide deposit depletion effects inside this layer are to be taken into consideration.

The availability of hydroxyl ions is described by:

$$N(OH^-) = CD\,(1\text{-}CE)/F \tag{1}$$

where $N(OH^-)$ is the flux of hydroxyl ions in mol/m^2/sec; CD is the current density in kA/m^2; CE is the current efficiency with regard to the production of caustic; F = Faraday constant.

The actual pH gradient from anolyte to catholyte and the pH at the surface in particular is the result of these two fluxes. Figures 5A and 5B give artist's impressions of the effects mentioned above.

9 SUMMARY OF FACTORS INFLUENCING THE BEHAVIOR OF IRON IONS

9.1 Current Efficiency

If due to the choice of the membrane, or the choice of anolyte strength, or for example impurity-aging of the membrane, current efficiency should decrease then the flux of hydroxyl ions will increase. This causes, all other factors constant, a shift of the pH gradient towards the anolyte; and the pH at the surface will rise.

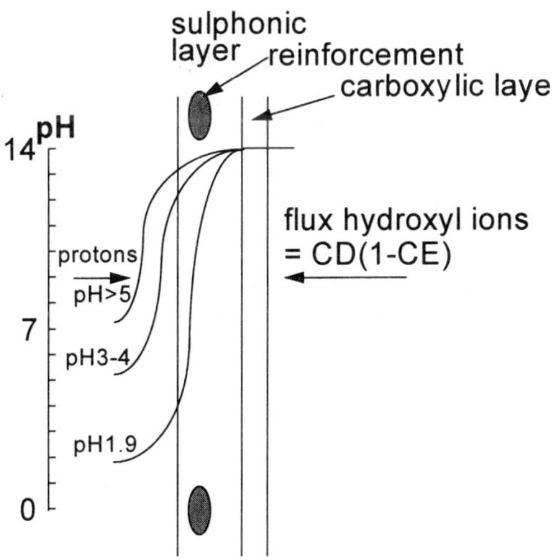

Figure 5A *The acidity of anolyte: a major factor that influences the pH gradient and the pH at the interface.*

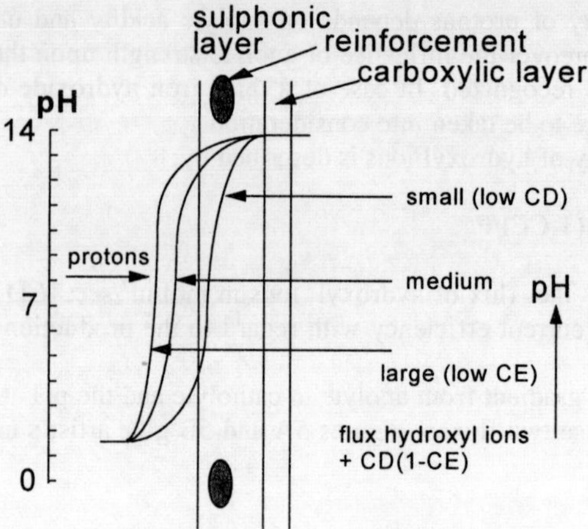

Figure 5B *The hydroxyl ions that are passing through the membrane i.e., the other factor that influences pH at the interface and the pH gradient.*

9.2 Current Density

Upon a decrease in current density the flux of hydroxyl ions will decrease and the pH gradient will move towards the catholyte. The pH at the anolyte membrane surface will decrease; and in the case of acidified brine the risk of iron ions entering the membrane will increase.

9.3 Caustic Strength

Upon an increase in catholyte strength more hydroxyl ions will be available, the membrane will adjust its equilibria to this increased level of hydroxyl ions. If current efficiency should decrease then the pH gradient will move towards the anolyte. pH at the anolyte surface of the membrane will go up.

9.4 Anolyte Acidity

If, due to increasing acidity of the anolyte, more protons are available then the gradient will move into the interior of the membrane. The pH at the membrane surface will decrease.

9.5 Anolyte Strength

Anolyte strength will influence current efficiency; moreover it will affect the activity of protons.

9.6 Water Transport

The water transferred through the membrane is the carrier for the transport of all other cations besides sodium. The water transport therefore codetermines the transport of the cations.

9.7 Resume

There is a series of factors which influence the precipitation of iron outside or inside the membrane. Neglecting these factors up to now might well account for the controversies regarding the effect of iron.

10 AN ANALYSIS OF MODERN CELL DESIGNS WITH REGARD TO THE FACTORS THAT GOVERN THE pH NEAR THE MEMBRANE SURFACE IN INDUSTRIAL ELECTROLYSERS

An industrial electrolyser will contain:

- if it is a bipolar electrolyser 1-2.7 m^2 of membrane per anolyte/catholyte compartment.
- if it is a monopolar electrolyser 20 - 40 m^2 of membrane per anolyte/catholyte compartment.

The larger the membrane surface area the more difficult it will be for a certain parameter, out of the list mentioned in previous paragraph, to be controlled all over the membrane area. With respect to obtain iron hydroxide precipitates outside the membrane it was shown to be desirable to control pH at the membrane/anolyte interface to be above 5 and below 11. This requires control of current density, of anolyte and catholyte strength and of acidity of anolyte. The recommendation and the areas of operation that should be avoided are depicted in Figure 6. Fortunately, the following can be observed:

(a) all over the world more new investments are made in bipolar designs than in monopolar designs. There is thus a trend towards smaller surface areas.

(b) baffles and other means of enhancing internal circulations and internal mixing are introduced in all designs. They are in fact a common feature in modern cell design.

These trends promote a more uniform anolyte strength, a more uniform catholyte strength, and a more uniform pH all over the membrane surface.

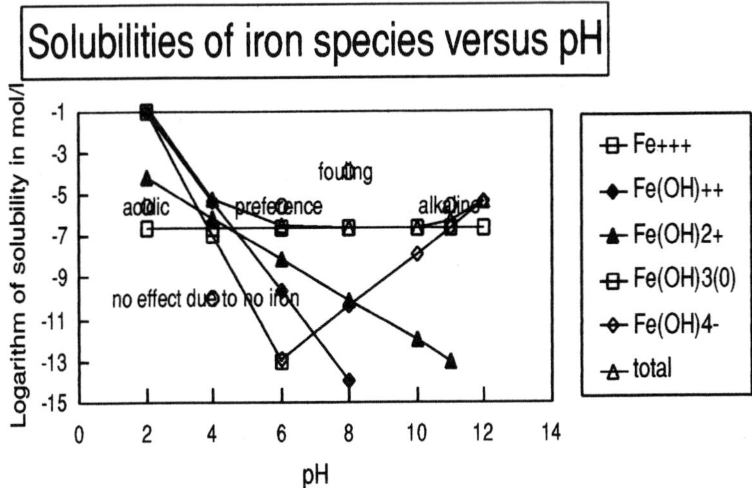

Figure 6 *The preferred area of operation and areas that should be avoided.*

11 CONCLUSION

(a) Feed brine with less than 5 ppb iron carries insufficient amounts of iron to cause any effect. After four years of operation the membranes from those plants do not contain or carry any iron. The possible effect of iron can be perfectly solved by reducing its level in feed brine.

(b) If iron is present, conditions of minimum effect can be selected. For such conditions of minimum effect the iron hydroxide should precipitate outside the membrane. In order to obtain a pH at the surface, such that iron hydroxide precipitates on the outer surface of the membrane, the fluxes of protons and hydroxyl ions should be balanced. It is obvious that the membrane itself, as it controls the hydroxyl ions passing from catholyte to anolyte, is an important factor in balancing protons to hydroxyl ions.

On these considerations an area of preference in the solubility versus pH diagram is defined. This area is to be seen in the center of Figure 6. The pH at the membrane surface should be in the range of 5 to 10. In that case iron hydroxide will precipitate at the surface and the iron ions cannot penetrate into the membrane.

If the pH at the surface decreases to below 5, iron cations will penetrate into the membrane as shown by Naoki *et al.*[4]

If the pH at the surface exceeds 11, which may occur in the case of low current efficiency membranes or very thick iron hydroxide precipitates then iron anions will slowly start to penetrate into the membrane.

(c) Feed brine containing more than 100-200 ppb of iron will cause fouling, the extent depending on cell design. In case of a power shut down the fouling layer is eliminated. This explains the observed decrease of cell voltage that occurred after a power shut down during the industrial tests.

(d) In those cases where the anticaking agent from the feedstock is the main source of iron, decomposition of the complex and filtering of the resulting iron hydroxide is a proven technology.

References

1. Du Pont Polymers Brochure 219233A, Nafion Users Guide 1991.
2. Asahi Chemical Industry Co., Ltd, Aciplex users instructions.
3. Asahi Glass Co., Ltd, Flemion users instructions.
4. Y. Naoki, H. Takei, H. Hohda and H. Shiroki, Denki Kagaku Kyokai (1994). 18th Soda Industry Technology symposium.
5. H. Shiroki, Y. Noaki, M. Katayose and A. Kashiwada, 'Modern Chlor alkali Technology, Volume 6', Ed. R.W. Curry, The Royal Society of Chemistry, London, 1995, 222-233.
6. H.M.B. Gerner and R.D. Theobald, 'Modern Chlor alkali Technology, Volume 6', Ed. R.W. Curry, The Royal Society of Chemistry, London, 1995, 173-184.
7. Y. Ogata, S. Uchiyama, M. Hayashi, M. Yasuda and F. Hine, *J. Applied Electrochemistry*, 1990, **20**, 555-558.
8. R.T. McAndrew, S.S. Wang and W.R. Brown 'Hydrometallurgy CIM Bulletin', 1975, **68(753)**, 101-110.
9. Y. Ogata, T. Kojima, S. Uchiyama, M. Yasuda and F. Hine, *J. Electrochem. Soc.*, 1989, **136(1)**, 91-95.
10. T. Momose, N. Higuchi, O. Arimoto and K. Yamaguchi, *J. Electrochem. Soc.*, 1991, **138**, 735-741.

16
HIGH CURRENT DENSITY OPERATION – THE BEHAVIOR OF ION EXCHANGE MEMBRANES IN CHLORALKALI ELECTROLYZERS

James T. Keating and Hans M. B. Gerner

E.I. du Pont de Nemours & Co., Inc.
Nafion* Research Laboratories and Customer Service Laboratories
Wilmington, Delaware and Fayetteville, North Carolina
USA

1 INTRODUCTION

Over the last ten years membrane chloralkali technology has had market growth of over 10%/year. The technology will account for more than one-third of total chloralkali production by the turn of the century. With this growth has come experience and confidence on the part of chloralkali producers and corresponding demand for innovations and improvements in electrolyzers and especially membranes. Operation at higher current density is attractive as a way to meet increases in market demand, and to take advantage of energy rates that vary with time of day, and by season. Membrane technology has always been distinguished by the ease and speed with which current density can be varied. It is natural that users should want to extend the limits of current density.

This paper addresses these questions about current density:

1. Is membrane life shortened by high current density operation?
2. How is membrane performance affected by high current density?
3. What is the upper limit of current density for membranes?

2 MEMBRANE LIFE

There are two aspects to the question of membrane life. First, how much chlorine and caustic can a membrane make before wearing out, or in other words, how much current in the form of sodium ions and attendant water can a unit area of membrane transport before it fails by losing conductivity, selectivity, or physical strength? Second, does the rate at which the membrane carries current affect membrane life?

Commercial experience gives the answer to the first question. Nafion® 954 that had been in service for nine years was analyzed at our laboratory for cell performance and physical properties. Voltage was typical of new N954 and current efficiency had declined 2%. Physical properties are shown in Table 1 and compared to recently made N954. There

**Nafion is DuPont's registered trademark for its perfluorinated membranes*

Table 1 *Physical Strength of Membrane After Nine Years' Operation*

Nafion® 954	Tensile (kg/cm)	Force Constant (kg/cm)	Elongation (%)	Tear (g)
9 Years				
MD	4.3	69	12	2600
TD	4.5	54	10	2100
Unused				
MD	4.6	84	13	2800
TD	5.0	54	15	2350

MD = machine direction; TD = transverse direction

is no significant loss of physical strength. In nine years' service this membrane operated at 1.6-3.8 kA/m^2. If we estimate an average of 2.7 kA/m^2 for nine years, the membrane has run for 24 kA-years, and is by no means at the end of its useful life. Four years at 6 kA/m^2 would be comparable service, considering only the mass of material – sodium ions plus water – transported. Therefore leaving aside for the moment the question of how the rate of transport might affect the membrane, there can be no doubt that transport of great masses of material and long membrane life are compatible. There is no obvious limit to the lifetime of perfluorinated membranes in chloralkali service. The oxidative stability of the polymer and reinforcement is so well demonstrated as to be beyond question. Passage of ions and water do not change the polymer. So as long as there is no mechanical damage from abrasion and no physical damage from precipitates – brine impurities are the cause of these – or misoperation, the membrane should maintain good performance indefinitely.

As for the effect of the rate of current passage on membrane life, commercial experience has extended to 5kA/m^2 operation with no sign of shortened membrane life. There will be some upper limit at which the mass flow exceeds the carrying capacity of the membrane and failure occurs through overheating or overpressure. Experimentation at higher current densities is aimed at identifying problems and designing membranes to overcome them.

Raising current density will have predictable effects:

1. Increased anolyte depletion in the region closest to the membrane surface and increased water transport.
2. Greater catholyte concentration in the region closest to the membrane surface.
3. Increased heat load on the membrane because of resistance heating. Internal pressure will also be a factor as higher temperature leads to greater vapor pressure within the membrane. In sulfonic/carboxylic bimembranes, which are the majority of commercial membranes used in production of 30-35% caustic soda, this effect will be greater in the carboxyl layer of the membrane, which is less conductive.

3 LABORATORY TESTING

Experimentation was done in circular laboratory cells of 45 cm^2 area, with ~500 cm^3 anode compartments and ~50 cm^3 cathode compartments. This small cell has many limitations compared to commercial cells, and the laboratory tests exaggerate problems arising at high current density. Since the objective was to identify changes at high current density, this is not a disadvantage. However, phenomena seen at 6 or 8 kA/m^2 in this small electrolyzer may not be found at less than 8 or 10 kA/m^2 in commercial electrolyzers. Sulfonic/carboxylic bimembranes were used in the electrolysis of sodium chloride in all but one of the experiments.

3.1 Voltage and Current Density

The first experiment determined the response of voltage to current density. The result is shown in Figure 1. Up to 8 kA/m^2, voltage is a linear function of current density, as it should be if resistances of the electrolyte and membrane are unchanged. Above 8 kA/m^2 curvature is seen, probably reflecting an increase in the gas fraction in the liquid electrolytes, and perhaps changes in the membrane. The result shows that testing at least up to 8 kA/m^2 is possible in the small laboratory cell.

3.2 Current Cycling 3 $\rightarrow$ 6 $\rightarrow$ 8 $\rightarrow$ 3kA/m^2

Three membranes of different polymer type and construction were cycled from normal to high to very high current density. Figure 2 shows the effect on current efficiency. The membranes differ in current efficiency, but all are above 95% and perform well at 6 kA/m^2.

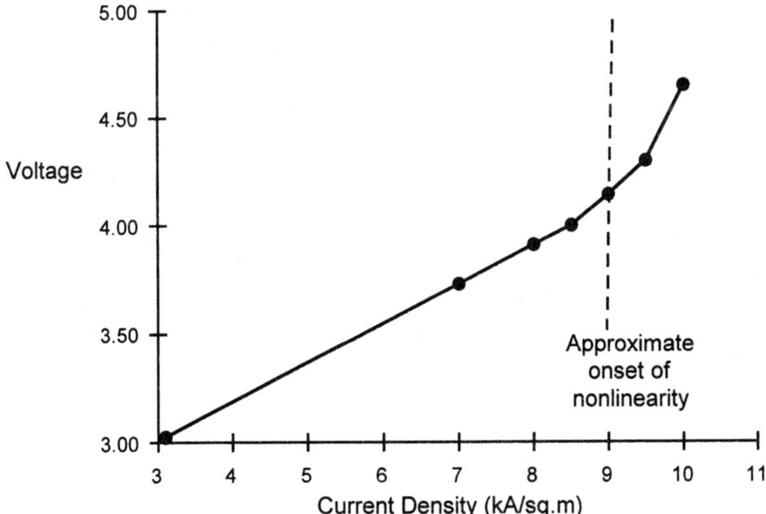

Figure 1 *Voltage versus current density in 45cm² test electrolyzer*

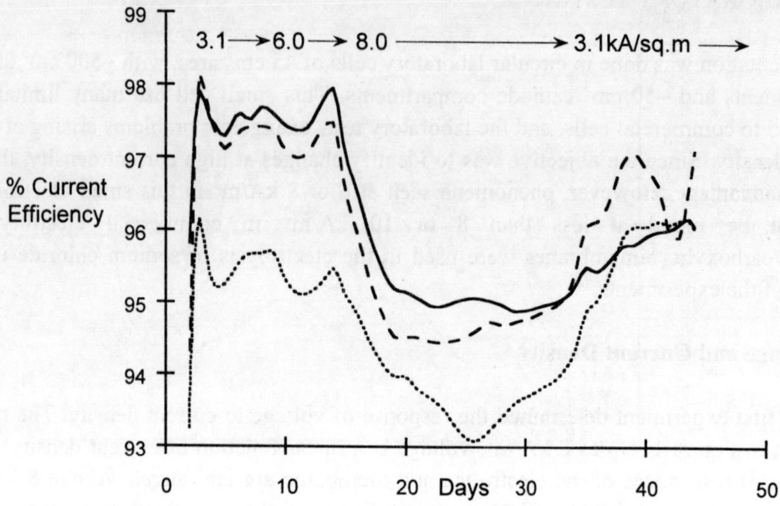

Figure 2 *Current efficiency as a function of current density*

At 8 kA/m^2 all drop ~2% in current efficiency. Those that were 97+% remain >95% at 8 kA/m^2, but the membrane that had just over 95% current efficiency at 6 kA/m^2 fell to 93%. Membrane performance began to recover when the current density was returned to 3.1 kA/m^2.

Figure 3 shows the response of water transport to current density. Comparing Figures 2 and 3, it is apparent that the change in current efficiency reflects the change in water transport. The membrane with the highest water transport has the lowest current efficiency. The decrease in current efficiency and increase in water transport indicate that the membrane is swelling under the pressure of mass flow at higher current density. Swollen membrane passes more water from the anolyte and more water is available because of anolyte depletion at the membrane surface. The swollen membrane also permits more leakage of hydroxide ions from the catholyte to the anolyte, reducing current efficiency. The fact that the membranes respond differently means that polymer properties and membrane construction influence elasticity and thus the degree of swelling. That the swelling became apparent only at 8 kA/m^2 may mean that either there is a threshold below which the membrane does not change in response to current density, or that the change occurs faster at higher current density. If the 6 kA/m^2 portion of the test were prolonged, a decline in current efficiency and increase in water transport might occur. This possibility was tested next.

3.3 Two-hundred Day Test at 7 kA/m^2

A longer test was run at 7 kA/m^2 to see how time influences performance, i.e. would a steady decline in current efficiency occur. Figure 4 shows the results using Nafion® 981 in a 200 day test.

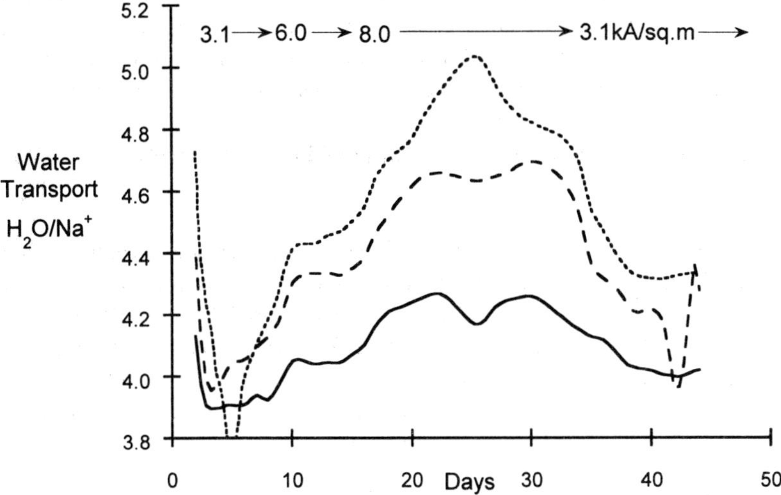

Figure 3 *Water transport (H_2O/Na^+) as a function of current density*

Time dependence of current efficiency is seen. Current efficiency starts about 97% and declines to ~95% over 100 days, leveling out after that. Voltage shows some decline in the course of the test. This may be related to the current efficiency decline since membrane conductivity can reasonably be expected to increase as anion rejection decreases. There seems to be no decline after 100 days, i.e. no continuous decline in current efficiency.

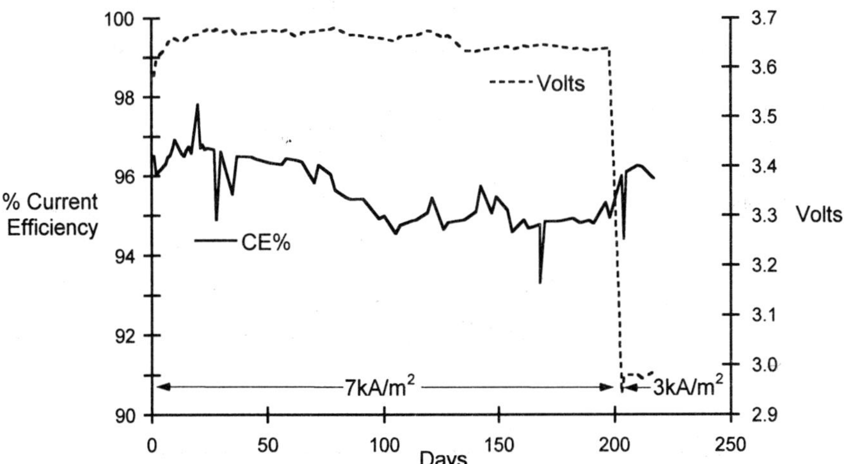

Figure 4 *Nafion® 981 at 7kA/m², 200 days, 45cm² laboratory electrolyzer*

After 200 days, current density was reduced to 3.1 kA/m^2. Voltage fell at once to about 2.95V, which is typical for this membrane in this cell under these conditions. Current efficiency began to rise. This reversibility is consistent with the 8 kA/m^2 test and is evidence that elasticity is maintained even after six months at high current density.

3.4 Behavior of Sulfonic Membrane in Potassium Chloride Electrolysis at 3.1 → 5 →3.1 kA/m^2

Figure 5 shows the variation of current efficiency with caustic strength at 3.1 and 5 kA/m^2 in the electrolysis of potassium chloride using an all-sulfonic membrane. Current efficiency at 5 kA/m^2 is >95% at ~25%KOH, but declines to 91% when KOH is increased to 32%. At 3.1 kA/m^2 current efficiency is >95% at 32% KOH. The KCl electrolysis shows greater sensitivity to current density and caustic strength than was seen in the case of NaCl. The difference may be related to the lower water transport that is typical of KCl electrolysis and lower water content of the membrane.

3.5 Physical Changes in Membrane at 10 kA/m^2

The current density described in the above work is calculated from the power supply amperage and the electrode dimensions. However, current density within the membrane, internal current density, is the critical factor for performance. The maximum local current density, not the average, will determine the point at which the greatest effect will be seen. To test the influence of internal current distribution, three membranes were made.

- Membrane **A** had regions of different conductivity. In operation, some parts would receive more than the average current density, some less, with a wide range of variability.

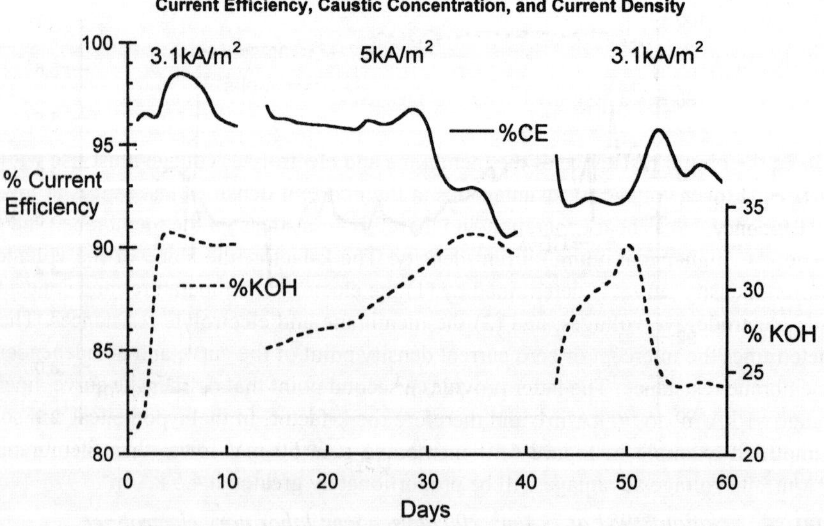

Figure 5 *Electrolysis of potassium chloride using a sulfonic membrane*

- Membrane **B** was reinforced with fabric that obstructed 20-30% of the current flow. The fraction of the membrane shielded by the fabric would receive little current and the rest more than average. These two regions were more or less uniform and variability is not so great as in the case of **A**.
- Membrane **C** was similar to **B** but more lightly reinforced so the maximum current density experienced in any part is less than in **B**.

The membranes were operated at 10 kA/m^2 for two months. Current efficiency averaged 93.6% for **A**; 94.1% for **B**, and 94.6% for **C**. Except for membrane **A**, voltage was stable. At the end of the experiment, the membranes were inspected for physical changes. Wrinkling was observed in **A**, less in **B**, almost none in **C**. Scanning electron microscopy (SEM) of the cathode (carboxyl) surfaces of the membranes showed differences in appearance. (Figure 5)

Membrane **A** has submicron diameter holes in the surface, marking regions of localized excessively high current density. The effect is similar to that seen when anolyte is over-acidified. In both cases the proximate cause is the same: excessive current load. In the present case, the local current density is too high; in the case of over-acidification, the membrane looses conductivity when the carboxyl groups are protonated. Membrane **B** has eruptions indicating a general overloading. Membrane **C** is less affected, with few eruptions seen.

The experiment confirms the importance of internal current distribution and shows the consequences of local overloading. The robustness of the membranes is demonstrated by the stable current efficiencies of all the membranes despite the changes they were undergoing. No doubt, longer operation would have affected performance, but it seems likely that brief exposure to excessive current density could be tolerated. Furthermore, where load shedding is practiced and the high current density phases last hours rather than days, the allowable upper current density limits may be greater than would be the case when operation is continuous.

4 VOLTAGE VERSUS CURRENT DENSITY

Given the ohmic resistance of the membrane and electrolyte, voltage must rise with current density. Lower voltage is advantageous at high current densities just as it is at low. Lower voltage means less heat generation and therefore less stress on the membrane. The ultimate benefit is higher maximum current density. The k-factor, the slope of the voltage versus current density curve, is determined by (1) the electrolyte temperatures and concentrations, and electrode overvoltages, and (2) the membrane and electrolyte resistances. The former determines the intercept or zero current density point of the curve and is independent of the membrane resistance. The latter provides a second point that defines the curve, linear in the range ~1 kA/m^2 to >6 kA/m^2, and therefore the k-factor. In the hypothetical and somewhat simplified example in Figure 6, if membrane α is 50 mV lower than membrane β at 3 kA/m^2, its voltage advantage will be proportionately greater at 4.5 kA/m^2.

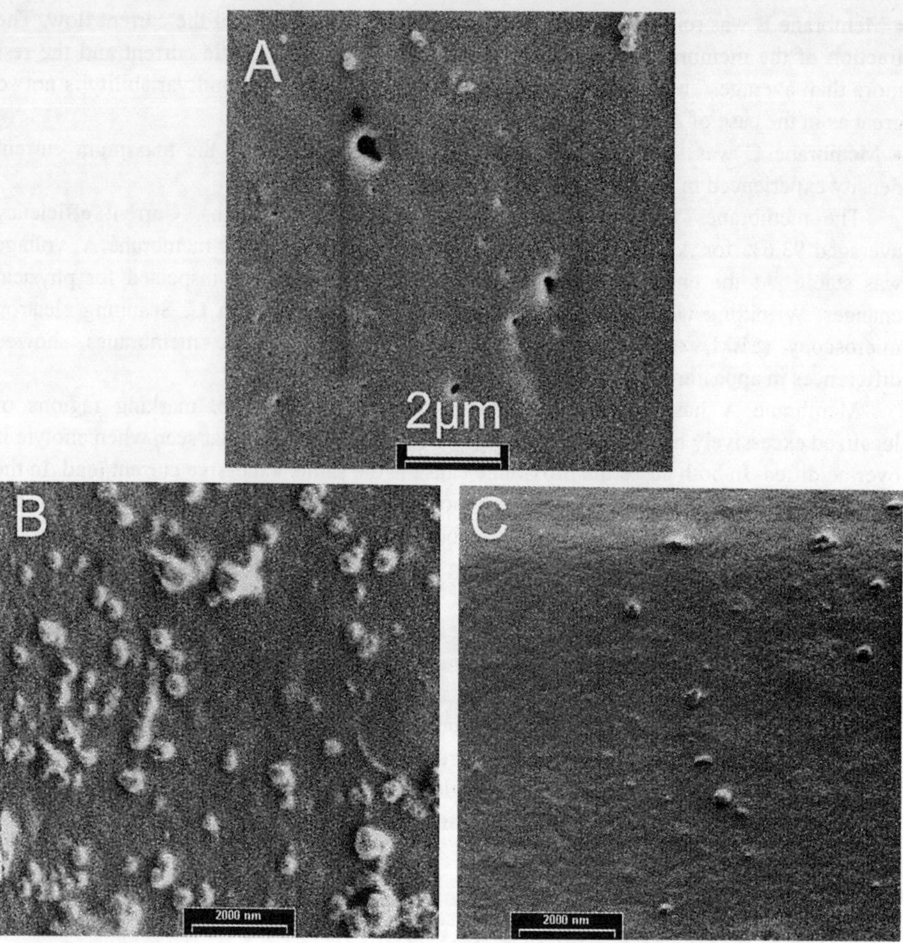

Figure 5 *Scanning electron microscopic photographs of three membranes with different internal current distributions after two months' operation at 10 kA/m²*

5 SUMMARY

Three questions were posed at the beginning of this paper. They cannot be answered categorically but this much can be said:

1. The long lifetimes found in membranes operated up to 4 kA/m² indicate that there is no apparent limit to how much caustic can be produced by a unit area of membrane if impurities and operating conditions are controlled.

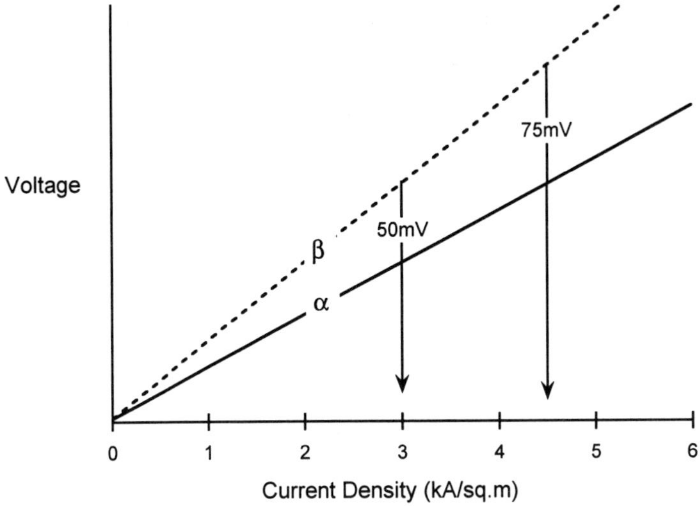

Figure 6 *Voltage as a function of current density*

2. Increasing current density reduces current efficiency. In the 45 cm^2 electrolyzer, ~2% decline is seen at 7-8 kA/m^2. In commercial electrolyzers with their better circulation and gas handling, this effect may be smaller.
3. Continuous operation at 6 kA/m^2 seems feasible based on this laboratory work. Intermittent operation at higher current densities should be possible.

Because of differences in electrolyzers, operating conditions, and brine quality, the decision to move to higher current density should be made in consultation with the membrane and electrolyzer manufacturers. It would be prudent to increase current density gradually, allowing time between changes to observe performance, particularly changes in water transport and the necessary adjustments in brine flow and catholyte dilution water. Operators may also need time to get used to the new conditions. Raising current density increases the rate of production but also the rate at which damage can occur from upsets in brine purification, acidification, control of temperature and concentration, and electrolyte feed rates.

The research reported here shows that no radical redesign of membranes is necessary for high current density applications. Characteristics that are recognized as beneficial for membranes today will increase in importance as current density is raised. Low voltage is more critical at 6 kA/mV than at 3 kA/mV, where the heat load is less and differences among membrane are less magnified. Good current efficiency will be more necessary than ever to counteract the effect of high current density. Current distribution in the membrane becomes more critical to prevent local overloading and consequent damage. Membrane surfaces that promote mixing and gas release have always been desirable and become more so at higher current densities. Recognition of these factors is guiding innovations in membrane design.

6 ACKNOWLEDGMENT

We are grateful to R. Altman, M.K. Gordon, L. John, J.B. Landau, M.D. Matthews, G.L. McCauley, H. Moore, and D.L. Sheppard for carrying out the experimental work reported on here, and to R.H. Senigo for the SEM photography.

17

PHASED CONVERSION OF A DIAPHRAGM PLANT TO MEMBRANE TECHNOLOGY

K. Stanley

ICI Chemicals &Polymers Ltd
PO Box 14
The Heath
Runcorn Cheshire WA7 4QF

1 INTRODUCTION

It has been thought for some time that the production of chlorine using mercury cells would be seen as an obvious candidate for conversion to membrane electrolysers. Indeed, this has been the case following the conversion of mercury cells to membrane electrolysers for the Japanese chloralkali industry for the production of sodium hydroxide. The European chloralkali industry is fundamentally based on mercury cells (around 2/3 of the production is from mercury cells) and celebrates its 100th operating year this year following the start up of the world's first mercury cell room in 1897 at ICI's Runcorn, England plant. Environmental pressure has been strong in opposition to mercury technology but producers have met and exceeded strict legislation so much so that today effluent streams from some chloralkali production sites are cleaner than the incoming water streams. Most people associate the year 2010 with the phase out of chloralkali production using mercury technology in Europe which was a voluntary target set by governments and industry.

From the outcome of several recent marketing studies it has been suggested that not all of the existing mercury electrolyser production will be converted, (in fact some studies show only perhaps 60% will be converted with the remainder shut down). This naturally would lead to a complete imbalance in the chlorine and caustic soda markets with European chlorine priced higher than now to offset new plant costs and a short fall of around 2 million tonnes of caustic soda in Europe. Assuming this shortfall is met from North America then freight charges would increase the pricing together with a short supply. It would however be an answer to most North American producers prayers with EDC being shipped to Asia and caustic to Europe.

So whilst the spot light has been placed on the mercury cells over the past few years the diaphragm cells around the world have been steadily producing product, but not without some questioning.

Diaphragm cells, which are the predominant technology in North America, are starting to feel some pressure from both outside and within the production facility.

Environmental pressure against the use of asbestos in general has impacted the chloralkali industry even though the use of asbestos in this industry is a minor percentage of the total use of asbestos in the world.

The majority of diaphragm chloralkali plants in the world are over twenty years old and as such, are starting to feel the need for injections of capital for overhauls and maintenance. This may not be the best answer as there is a very viable alternative which is phased conversions to membrane electrolysers.

2 MAINTENANCE OF DIAPHRAGM CELLS

Diaphragm cells produce a weak caustic soda solution in brine and consequently require multiple effect caustic evaporators and salt separators to concentrate this caustic soda solution to 50% strength. Even so, the 50% caustic soda solution still contains a high level of sodium chloride left over from the brine solution. The evaporators require a high stream usage to produce 50% caustic soda from an original 12-14% caustic soda in brine solution and are a high maintenance item especially when levels of sodium hypochlorite exceed normal levels due to losses of efficiency of the diaphragm cells. This leads to increased evaporator corrosion and more frequent downtime.

Diaphragm cells work best at full steady load and do not take kindly to either shutdowns or frequent changes in load which can dramatically affect their current efficiency and can lead to the possibility of rediaphragming all the cells.

Recently, sources of asbestos have become more and more of a problem. Changes in supplies of asbestos require some changes in the way diaphragms are deposited mainly due to the different mix of fibres and trials need to be undertaken to achieve the highest operating cell efficiency with any new source of raw materials. All of this gives an operating plant more work and more costs than before. You may well say that the answer to this particular problem lies in synthetic diaphragms which eliminate the need for asbestos and provide a longer lasting reproducible diaphragm. There are several synthetic diaphragm raw material producers offering a package of material and know how to the market but there seems to be one fundamental flaw to this and that is that the diaphragm cells are twenty or more years old. With anything that is that old and particularly something that manufactures chlorine and caustic soda age starts to have a large effect on the equipment.

If we look at a diaphragm cell it is made up of four distinct parts; a base, anodes, a cathode box and a hood or cover.

Bases are usually made of steel protected inside the cell by either a rubber mat or a titanium sheet to prevent corrosion by the chlorinated brine. Outside of the cell the bases also perform the function of transferring electricity to the anodes so copper is usually bolted to the steel (or sometimes explosion bonded). Conditions outside of the working environment usually consist of drips of either brine or weak caustic soda solution onto these electrical contacts.

Anodes in diaphragm cells enjoy the advantage of the fact that out of the three main chlorine cell types (diaphragm, mercury, membrane) diaphragm cell operating conditions are the least aggressive to anode coatings. This leads to lifetimes of fifteen to twenty years for diaphragm anode coatings.

Cathode boxes, apart from providing a substrate for the diaphragm to be deposited onto also incorporate the steel mesh for the cathodic side of the electrolyser. As the diaphragm is porous to all liquids all of the depleted brine passes through the diaphragm to the cathode. Brine containing some chlorine also passes through and the chlorine is reacted eventually to sodium hypochlorite. The steel mesh of the cathode boxes undergoes some corrosion over the years due to chemical attack and periodically requires a new mesh. Eventually the mesh and current collectors or "fingers" are corroded sufficiently to require replacement.

Diaphragm cell covers or hoods are fabricated from many different materials such as FRP, FRP lined concrete, titanium, plastics and composites. All have the role of keeping the chlorine and brine in the cell. Again they each have a finite lifetime and eventually require replacement. I have not mentioned the fact that periodic replacement of the asbestos diaphragms, usually associated with the replacement of the gasketing systems, is required and this frequency in time is dependent on the efficiency of the diaphragm and of course the flow rate through the diaphragm or brine head.

If we look at all of these aspects coming together at around the twenty year lifetime one can see a large refurbishment programme and cost.

To renew a base, cathode box and hood is a straight forward decision based on wear and tear and cost but what about the anode coatings.

A large proportion of anodes and anode coatings are leased whilst others were purchased outright at the initial sale. For the purchased electrodes is it wise to recoat with a product that will last another fifteen to twenty years if the cell around it may not stand the ravages of time? For leased electrodes the decision is somewhat simpler in that the coating is normally based on a price per tonne produced. If they are not needed they are simply returned.

3 PHASED CONVERSION

Taking each of these aspects into account from environmental, quality of product through to refurbishment costs it is now becoming economically sensible to look at the phased conversion of diaphragm cells to membrane electrolysers.

The phased conversion of a diaphragm plant enables the costs to be spread over a number of years rather than a single high capital project.

Also most of the diaphragm plant equipment continues to be used.

Membrane electrolyser conversions of diaphragm plants are relatively simple providing some basic fundamental rules are followed.

In a diaphragm plant, recovered salt from the multi effect evaporation unit is returned into the feed brine circuit to strengthen the brine. Whilst this seems a very basic unit operation to perform it has a bearing on the performance of the membranes in membrane electrolysers when a phased diaphragm plant conversion is undertaken. Salt from an evaporator usually contains some metals (as a corrosion product from the steel of the evaporators), the most common being nickel and chromium. As the recovered salt strengthens the brine circuit these metals could end up on the brine side (or anode side) of the membrane and try as hard as they can to get through the membrane to the cathode side. Many of us have seen the membrane suppliers specifications over the last few years reducing and reducing as far as allowable limits of metals in brine are concerned, particularly nickel, as it is very efficient in causing an increase in resistance and hence voltage in the membrane. The obvious thing to do is to eliminate the metals in the feed brine by adding the recovered salt to the weak brine from the membrane plant and using it as feed brine only to the diaphragm plant in a phased conversion. Eventually, there will be no recovered salt as the phased conversion is completed.

Steam usage for the multi effect evaporators reduces as the phased conversion progresses since less steam is required to evaporate from 32% to 50% caustic soda. Eventually the salt recovery plant will be eliminated completely and the days of vibrating centrifuges will be over. Interestingly since there is less salt to recover from a similar volume of liquid, cooling capacity has to be increased to crystallise the salt during the phased conversion.

ICI has developed a new electrolyser which is particularly suitable for the phased conversion of diaphragm cell rooms. The ICI membrane electrolyser, called the FM1500, accommodates all of the latest developments including circulation enhancers to ensure higher current efficiencies. Indeed, its design enables it to be retrofitted to the majority of diaphragm plants in the world to accomplish phased conversions.

ICI's FM1500 membrane electrolysers take up less space than their equivalent diaphragm cells so more electrolysers can be physically installed in a cellroom.

Additionally due to their low operating voltage existing rectifiers can accommodate a larger number of membrane electrolysers. Usually the working area or gallery of a diaphragm plant is between rows of diaphragm electrolysers and is the same as that required for ICI's FM1500 membrane electrolysers. This means that access to all the diaphragm cells or membrane electrolysers is from a common operating area. Some piping can be reused for headers to enable gaseous products to be fed into the existing plant. In a phased conversion liquors need to be separated to feed the pure brine to the membrane electrolysers and remove the caustic soda.

4 EXTRA EQUIPMENT REQUIREMENTS

So what extra equipment is required in addition to that already in place in a diaphragm plant?

In the brine treatment plant for a diaphragm cellroom vessels and equipment are in place to reduce the total hardness of the brine to around 3 ppm hardness to feed to the diaphragm electrolysers.

Membrane electrolysers require hardness levels of 20 ppb so extra equipment is essential. Brine at 3 ppm hardness may be fed directly to an ion exchange unit to reduce the level to 20 ppb. As a precautionary measure a cartridge or polishing filter may be installed up stream of the ion exchange unit to cater for diaphragm brine plant upsets. Pure depleted brine exiting from the membrane plant contains an amount of absorbed chlorine which needs to be removed in a dechlorination unit. In order to recover the chlorine a vacuum system is preferred (to recover strong chlorine) rather than an air blown dechlorinator which produces weak chlorine. If saleable sodium hypochlorite is required an air blown dechlorinator would suffice. Following primary dechlorination, to levels of 20 ppm or less dissolved chlorine in brine, a question needs to be asked as to whether this is low enough to be sent to the diaphragm plant brine system. Points against this are chlorination of pipework, open topped resaturators and acid addition of the diaphragm electrolyser feed brine. Liberated chlorine may give a harmful smell so normally secondary chemical dechlorination of the membrane depleted brine is undertaken. Chemicals frequently used for this are hydrogen peroxide, sodium bisulphite or sodium thiosulphate.

On the caustic soda side of the electrolyser the 32% sodium hydroxide can either be fed to a separate system or directed to the second effect of the diaphragm plant caustic evaporator. Although this will blend the product caustic soda it is the cheaper option during conversion.

Recovered deionised water from the evaporators may be used in the membrane plant for use in the caustic soda circuit as a dilution water.

5 EXPERIENCE

So how does one start up a membrane plant as part of the same electrical circuit as a diaphragm plant. I suppose the answer is far easier than when it was all diaphragm cells. Operators can concentrate more on the reduced number of diaphragm cells looking at the brine heads and cathode box overflows as the membrane electrolyser circuit requires no physical adjustments when going on load. The membrane circuit can take any load that the diaphragm plant requires to tighten up the diaphragms so there are no restrictions placed on the plant from the membrane side.

Eventually, during a phased conversion, a decision will need to be taken to convert the remainder of the plant based on the water balance and salt recovery aspects.

This type of phased conversion has recently been undertaken at ICI Forest Products Bécancour plant in Quebec, Canada. It is undergoing a phased conversion of one of its diaphragm cellrooms to achieve a 200,000 T/y membrane plant. The plant was originally started in 1975 using Hooker H2A diaphragm cells. The economics of carrying on with the diaphragm plant or converting it to a membrane plant proved to be both economical and also market driven for a conversion decision. Once sanction was approved in May 1996 ICI's Electrochemical Technology Business based in the UK provided the basic design to the Bécancour team who contracted local companies to provide the detail design, procurement and erection up to mechanical completion.

The plant imposed a tight schedule for major equipment to be delivered in that it needed to be indoors and weather proofed by the beginning of November 1996. Winters in Quebec Province, Canada can be extremely cold with temperatures of -40 °C not uncommon. Items of equipment, such as valves, usually laid outside until required can be lost in the snow until spring. A plant shut down occurred in November 1996 to move a turnaround busbar so that the first phase of diaphragm cells could be removed. The floor of the cellroom area was cleared and then rebuilt with foundations for the membrane electrolysers including trenches. Once all of the erection was completed the FM1500 electrolysers were membraned and berthed in the cellroom. Membraning the FM1500 electrolysers is extremely simple and even easier than the previous ICI FM21 electrolysers, which were recognised in the industry as one of the easiest electrolysers to assemble and disassemble.

Design changes and developments incorporated into the design and manufacture of the FM1500s have enhanced their operating performance whilst simplifying their mechanical installation aspects. A second shut down enabled all of the new pipework for the membrane electrolysers to be tied in to the existing diaphragm systems and the turnaround busbar to be relocated back to its original position.

Start up of the cellroom circuit in April 1997, which now contained membrane electrolysers as well as diaphragm cells proved to be trouble free. Full load was achieved within two days with a performance test of the membrane plant being passed within the first week.

6 CONCLUSION

I have shown that phased conversions of diaphragm plants to membrane electrolysers is both technically and economically viable.

We now look forward to the next phase of the plant conversion in Canada.

We would be very happy to talk through this type of conversion with other interested operators and to explain the benefits of the ICI FM1500 membrane electrolysers.

18
IMPROVEMENTS OF BRINE QUALITY: REMOVAL OF BROMIDE, IODIDE AND CHLORATE

Alain Hanneuse

SOLVAY - General Technical Direction - Brussels (Belgium)

Luciano Chiti

SOLVAY - Electrolysis R&D Unit - Rosignano (Italy)

1 INTRODUCTION

The quality of brine has ever been a fundamental factor for a safe and stable operation of electrolysis processes (amalgam and diaphragm) and to get high quality products.

However, in the last decade, membrane technology and environmental considerations have increased the need for better brine quality.

Membrane technology introduced new requirements for brine quality to improve the stability of the performance of the membranes. Very low levels of calcium and magnesium have to be reached. Other components were discovered as harmful components for the membranes: strontium, barium, aluminium, sulphates, silica and also iodide.

Environmental protection requires improvement of waste water quality: the chlorate or bromate contents are concerned. Also, it requires minimization of waste production: halogenated organic by-products originated by bromine content of chlorine have to be reduced.

More and more, high quality products are needed: chlorine with very low content of bromine to avoid inseparable organic brominated components, caustic soda with very low content of chlorate to avoid corrosion of concentration units.

This paper presents recent experience of the Solvay group in the removal of three important impurities of the feed brine for chlor-alkali electrolysis: bromide, iodide and chlorates.

2 REMOVAL OF BROMIDE

Bromide is present in the salt, raw material of the electrolysis, in a very wide range of concentrations depending on the source of salt.

Bromide in brine produces bromine in the electrolytic cell that is mixed with the chlorine production. This bromine is a nuisance for most chlorine users producing organic chemicals (for example dichlorethane, chloromethanes, etc.) because it generates undesired by-products.

In some cases, it is very difficult to separate these brominated organics from the final product, in other cases they increase the quantity of halogenated organic wastes to be treated (Figure 1).

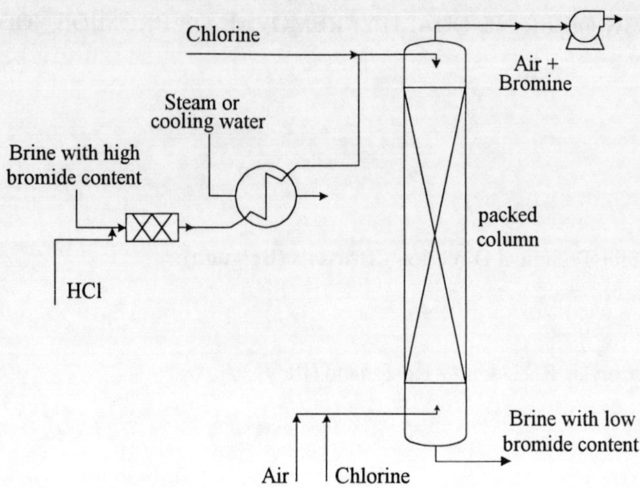

Figure 1 *Bromide removal process*

Several techniques are known and used to reduce the content of bromide in the brine with variable results and efficiencies normally below 95 %.

A considerable reduction of the level of bromide contained in NaCl or KCl brines can be achieved with a very well known technique: oxidizing of bromide to bromine by the action of chlorine added to the brine:

$$2\ Br^- + Cl_2 \rightarrow Br_2 + 2\ Cl^- \tag{1}$$

Bromine is removed by stripping with air.

A gaseous mix of air, bromine and traces of chlorine is separated and has to be treated.

Brine with very low bromide content is produced.

The correct operation of this process requires a good knowledge of a certain number of side-reactions:

Bromate formation:

$$Br_2 + H_2O \rightarrow HBrO + HBr \tag{2}$$
$$3\ HBrO \rightarrow HBrO_3 + 2\ HBr \tag{3}$$

Ammonium destruction:

$$Cl_2 + H_2O \rightarrow HClO + HCl \tag{4}$$
$$2\ NH_3 + 3ClO^- \rightarrow N_2 + 3\ Cl^- + 3\ H_2O \tag{5}$$
$$2\ NH_3 + 3\ Cl_2 \rightarrow N_2 + 6\ HCl \tag{6}$$

Chloramines formation:

$$NH_3 + Cl_2 \rightarrow NH_2Cl + HCl \tag{7}$$
$$NH_2Cl + Cl_2 \rightarrow NHCl_2 + HCl \tag{8}$$
$$NHCl_2 + Cl_2 \rightarrow NCl_3 + HCl \tag{9}$$

Desired reactions are those that lead to ammonium destruction but these reactions also consume chlorine. Other chlorine-consuming reactions are not desired like formation of bromates and chloramines.

The operations parameters have to be adapted to ensure a complete reaction of bromide to bromine, that can be easily stripped, and avoid the formation of bromates that remain in the brine.

The four operating parameters that have an influence are:

- pH (acid)
- Temperature (> 65 °C)
- Air flow
- Chlorine flow

Adapting those parameters, we reached the following results:

- At Rosignano (Italy), an industrial plant (150 m^3/h brine) has operated since 1990 producing brine containing less than 0.5 ppm Br from brine containing about 50 ppm Br.
- In various pilot trials with brines (NaCl and KCl) containing more than 100 ppm Br, we always succeeded in obtaining brines containing less than 0.5 ppm Br.

The efficiency was ever greater than 99 %.

In the case of lower initial bromide content, normally below 40 ppm, stripping with air can be substituted with ion-exchange resins with similar final results.

3 REMOVAL OF IODIDE

Iodide is present in the salt, raw material of the electrolysis.

Iodide in brine precipitates in the iodate form and causes loss of current efficiency of the membranes. A synergetic effect has been reported especially in combination with barium.

Therefore membrane suppliers require less than 0.4 (even 0.2 ppm) in the feed brine of membrane electrolysers.

Iodide must first be oxidized to iodine by addition of chlorine:

$$2\,I^- + Cl_2 \rightarrow I_2 + 2\,Cl^- \qquad (10)$$

Then iodine can be fixed by anionic exchange resins with halogenated groups:

$$I_2 + Cl^- \rightarrow (I_2Cl)^- \qquad (11)$$
$$R^+Cl^- + (I_2Cl)^- \rightarrow R^+(I_2Cl)^- + Cl^- \qquad (12)$$

where R^+ is a cationic fixed site of the resin.

The brine to be treated is acidified. Chlorine is produced in the brine pipe with an in-line electrolyser. The current of this electrolyser is adapted automatically to maintain a pre-defined redox potential at the output of this electrolyser and avoid the formation of undesired iodate (Figure 2).

Then iodine is fixed with anionic exchange resins.

To avoid the undesired production of iodate, it is important to regulate:

- pH (< 3)
- redox potential after chlorine addition (controlled actuating on an in-line electrolyser).

With this technology, the following results were obtained:

At Rosignano (Italy) the feed brine for membrane pilot cells with about 3 ppm iodide is treated since 1992 with this process. Iodide in the outlet is ever been below 0.2 ppm.

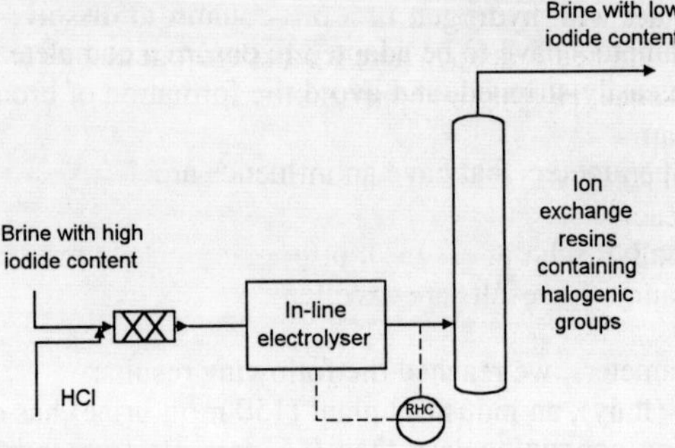

Figure 2 *Iodide removal process*

4 REMOVAL OF CHLORATE

Chlorate is generated in the anodic compartment of the membrane electrolyser by the migration of OH^- coming from the cathodic compartment.

High chlorate level in the brine loop increases the chlorate contamination of caustic soda and causes corrosion of the concentration unit.

High chlorate level in the brine purge can be a problem for waste water regulation.

Commonly, chlorates are removed by purging brine or treating a part of the brine loop by acidifying at high temperature. This process is consuming great quantities of acid and is consuming steam.

A new process named CCR (Chlorate Catalytic Reduction) was recently developed by our catalyst research unit (Figure 3).

Chlorate can be reduced to chloride by the action of hydrogen in the presence of a precious metal catalyst:

$$ClO_3^- + 3\,H_2 \xrightarrow{\text{precious metal catalyst}} Cl^- + 3\,H_2O \qquad (13)$$

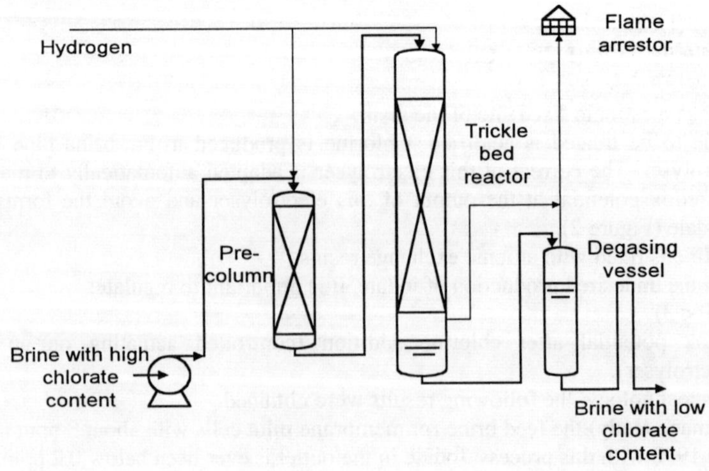

Figure 3 *Chlorate removal process*

Brine is put in contact with hydrogen in a pre-column to dissolve hydrogen in water. Reaction occurs in a trickle bed reactor where a gas (hydrogen) a liquid (brine) and a solid (the catalyst) are in presence. Hydrogen is then separated from the brine.

Essential elements are:

- Precious metal catalyst
- Three phase reactor

This process is actually tested in a 1 m³/h pilot plant in Jemeppe (Belgium) (Figure 4). After 2 months of operation the results are excellent.

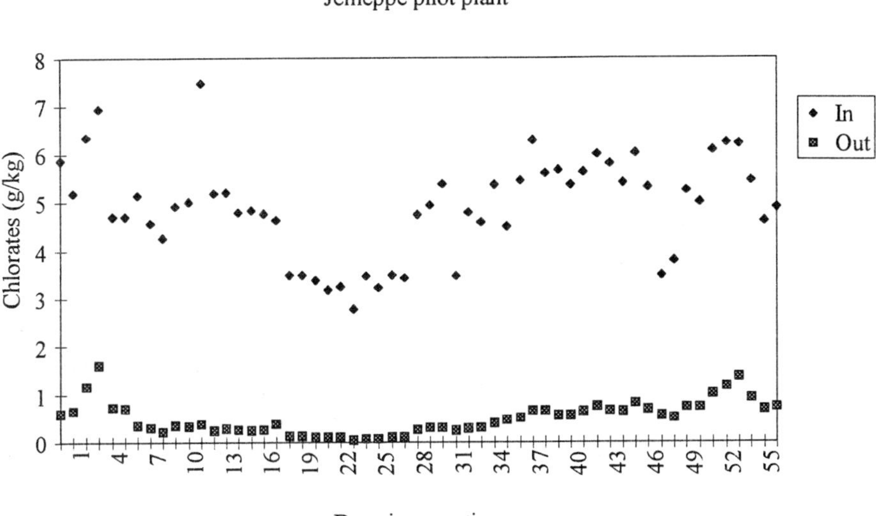

Figure 4 *Results of Jemeppe pilot plant*

Brine at the output has level of chlorates below 1 g/kg starting from brine with about 5 g/kg. With a partial flow installation, it should be possible to maintain the chlorate level in the brine below 5 g/kg or less.

Investment costs are in the same order as an acidification at high temperature but operating costs are significantly lower.

19

PPG'S TEPHRAM® DIAPHRAGM: THE ADAPTABLE NON-ASBESTOS DIAPHRAGM

Peter C. Foller, Donald W. DuBois and Joe Hutchins

PPG Industries, Inc.
Chemical Technical Center
440 College Park Drive
Monroeville, PA 15146 USA

1 INTRODUCTION

The chlor-alkali industry has concerns over the continued availability and acceptable utilization of asbestos. In response to these concerns, several firms have attempted the development of synthetic diaphragms. OxyTech's commercialization efforts are the most well known; [1,2] however, a publication by Rhone-Poulenc is also relevant. [3]

The goal for PPG's Tephram® diaphragm program is to commercialize a fluorocarbon-based replacement for asbestos diaphragms that uses depositing and operating procedures which, as closely as possible, mimic those of the asbestos diaphragms they replace. Realizing materials costs will be higher for any non-asbestos approach, subsidiary goals of the program are to achieve both longer diaphragm life and a power consumption advantage to help justify a higher initial cost.

As background, diaphragm chlor-alkali cells have anolyte and catholyte compartments that are separated from one another by a porous diaphragm. Brine at about 300-320 g/l is fed to the anolyte, wherein chlorine is evolved from dimensionally stable "DSA" anodes. (DSA anodes are composed of selected metal oxides, predominantly RuO_2, thermally deposited onto titanium expanded metal.)

Unreacted brine flows through the diaphragm to the catholyte, wherein a steel cathode evolves hydrogen and produces sodium hydroxide. The brine fed to the anolyte is supplied at a fast enough rate such that an anolyte-to-catholyte level differential ("level") is maintained between set limits which serves to isolate anodically evolved chlorine from cathodically evolved hydrogen.

Current is passed to provide approximately 50% depletion of the feed brine. "Cell liquor" overflows from the catholyte. Cell liquor is a mixture of the unreacted brine and the cathodic product, caustic soda. The caustic soda is later separated from the unreacted brine by multiple effect evaporation, which precipitates solid salt and produces 50% caustic soda.

Typical diaphragm chlor-alkali cells operate at 2.9 to 3.6 V per cell, depending on the current density of operation and mechanical design factors. Temperatures are 90-95 °C. Currents fed to industrial cells typically range from 30 kA to 150 kA. Current densities range from 1.4 to 2.8 kA/m^2. Power consumption for diaphragm cells averages 2,800-3,000 kWh/mT on a chlorine basis.

The development of non-asbestos diaphragms represents a significant technical challenge. Chlor-alkali diaphragms must prevent the back reaction of chlorine with hydroxide ion, which produces hypochlorite and chlorates, and reliably isolate chlorine from hydrogen to prevent explosive conditions. Successful diaphragms must be: (1) inert to oxidation by chlorine, hypochlorite, and chlorate, (2) inert to as low as pH 2 (when operation with acid brine feed is selected by a plant) and to strong base, (3) mechanically strong, (4) resistant to bubble abrasion, (5) highly ionically conductive, (6) resistant to entrapping gas, and (7) efficient in the utilization of current under a wide range of current density and brine flow conditions ("current efficiency" is the measure of the fraction of current passed that results in useful products).

Of all diaphragm cell components, asbestos diaphragms typically have the shortest life. They are gradually affected by the chemical attack of alkaline catholyte and acidic anolyte (when acid brine is used), and by the accumulation of "plugging" impurities from the feed brine (such as salts of Ca, Mg, Ba, Fe, Ni, Mn, Si, and Al) which cause a cell to go "high level". Unsteady operating conditions are also potentially damaging. As a result, current efficiency drops as the back-migration of hydroxide ion is gradually less efficiently suppressed. Asbestos diaphragm lifetime varies from plant to plant, but typically ranges from 9 to 16 months.

Development of the Tephram® non-asbestos diaphragm began with laboratory work in 1986 and has been evolving since.[4-12] Laboratory results and extensive plant tests convinced PPG to further demonstrate the technology by the gradual conversion of a 160 mT Cl_2/day full circuit of 160 Columbia N-6 cells located at Natrium, West Virginia. The changeover was completed in May, 1992. Four generations of Tephram diaphragms are presently operating in this circuit.

Within PPG, Tephram diaphragms are also installed on an experimental basis in two other circuits at Natrium (one composed of Columbia N-3 cells and one composed of OxyTech MDC-55 cells) and at Plant C in Lake Charles, Louisiana, a circuit composed of PPG (Glanor) bipolar V-1244s.

Beginning in September 1994, Tephram diaphragms also began undergoing trials in non-PPG plants and, thus, experience in additional cell designs is being gained. As of this date, six non-PPG plants have been involved.

This paper describes the diaphragm itself, the deposition process, plant operations, and preliminary economics. Data for the Generation 4 diaphragm is presented for the first time. This is an evolutionary improvement which shows higher current efficiency than previous generations in larger cell types such as the OxyTech MDC-55 and PPG's (Glanor) V-1244s.

2 THE COMPOSITION OF TEPHRAM® DIAPHRAGMS AND THEIR DEPOSITION

The procedures for the deposition of Tephram® diaphragms onto cathodes are essentially the same as those used in the deposition of asbestos diaphragms. The deposition procedures use the same type of equipment as is normally found in diaphragm cell chlor-alkali plants. Some additional equipment is, however, required for slurry preparation.

Both of the two different slurries used in the deposition of Tephram diaphragms are prepared from commercially available materials. The materials used are individually and collectively highly resistant to either acid or alkaline brine, chlorine, caustic soda, hypochlorite, and chlorate.

Tephram diaphragms are composed of three basic components: (1) a "basecoat", or base diaphragm vacuum deposited onto cathodes, (2) a "topcoat" vacuum deposited on top of the basecoat, and (3) "dopant" materials periodically added to the anolyte of cells, on an individual basis, to form what we believe is a gel within the diaphragm and to maintain the diaphragm.

First, the basecoat is vacuum deposited from a slurry onto a clean cathode in much the same manner as asbestos diaphragms are deposited. PPG has experience with deposition onto both woven steel screen and perforated steel plate. The basecoat is composed of:

- PTFE fluoropolymer fiber (which helps form the mat)
- PTFE microfibrils (which provide for suitable diaphragm porosity)
- Nafion® perfluorinated ion exchange resin (which imparts long-term wettability to the PTFE components)
- Other "sacrificial" materials (which dissolve and/or oxidize during cell start-up)

Second, using a separate deposition system, the topcoat is added by vacuum deposition from a slurry containing inorganic particulate materials. The topcoat is incorporated into, and becomes an integral part of, the basecoat diaphragm. The purpose of the topcoat is to adjust diaphragm permeability and uniformity.

The finished diaphragm is then oven dried at low temperature (not sintered).

Much of the equipment needed for the deposition of Tephram diaphragms is available in most diaphragm cell chlor-alkali plants. The equipment needed includes deposition tanks, vacuum tanks, filtrate tanks, and a drying oven. Additional equipment is needed for PTFE microfibrils manufacture, mixing the slurry, and for depositing the topcoat. The total investment in new equipment or building space is plant-specific and depends on cell type, the size of the plant, and the type and availability of existing deposition equipment. The capital requirement for a 500 mT Cl_2/day diaphragm plant would typically be about $500,000.

3 "DOPING" OF DIAPHRAGMS

Tephram diaphragms must separate hydrogen and chlorine gases as well as separate anolyte from catholyte. This separation must be reliably performed with low cell voltage, a low rate of hydroxide ion back-diffusion against the brine flow, and good current efficiency. The porosity of the basecoat and topcoat are such that, alone, they cannot perform these functions. Thus, a third and key element of the diaphragm is incorporated.

Both soluble and insoluble materials added to the anolyte of cells during start-up and intermittently thereafter (denoted as "dopants") form an improved barrier to liquid flow and impede the back-diffusion/electromigration of hydroxyl ions. We believe a gel layer is formed within the diaphragm. The use of dopants results in more uniform brine flow across the diaphragm, higher current efficiencies, and control of diaphragm permeability (resulting in the control of anolyte-to-catholyte level differential).

After start-up doping protocols, the diaphragm is maintained by adding additional dopant materials to the cells periodically (see Figure 1). The use of various types of doping materials allows the semi-independent adjustment of level and current efficiency. An advantage of such maintenance by doping is the ability to adapt the performance of Tephram® diaphragms to the individual characteristics of a circuit (such as its current density and brine quality).

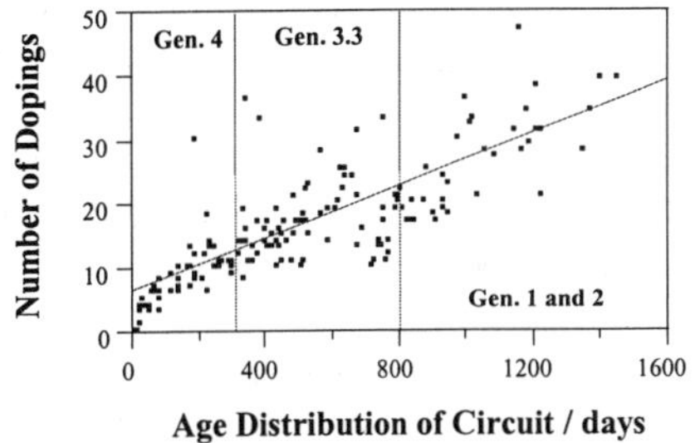

Natrium Cct. 5 Operating Conditions

(Applies to Figures 1 to 3)

- *Fully converted to Tephram® diaphragms*
- *30 kA normalization*
- *1.5 kA/m²*
- *120-130 gpl NaOH*

Figure 1 *The frequency of doping of individual cells at Natrium Circuit 5 versus time. The slope indicates doping each cell every 35-40 days.*

Further, as demonstrated in cells at Natrium, West Virginia, the use of dopant materials to adjust the diaphragm permeability has the added benefit of minimizing cut-outs for high anolyte-to-catholyte level, thus extending diaphragm life. If a cell develops a high level (perhaps due to accidental over-doping or brine quality upsets), it can recover over time if no further dopants are added.

4 LABORATORY EXPERIENCE

Favorable laboratory results in small (3" x 3") cells were achieved early in PPG's program. In these cells, the requirements of a successful diaphragm, as measured by cell power consumption, oxygen in chlorine, hydrogen in chlorine, chlorates in cell liquor, caustic concentration, and anolyte to catholyte level, were clearly met.

It rapidly emerged that a greater challenge lay in the translation of these laboratory results to plant practice. For example, the high current efficiencies (96+%) seen in the 3" x 3" cells were not reproduced in plant cells. To address this disconnect, several 30" tall cells simulating the hydraulics of plant cells were also installed in the laboratory. These cells are now successfully used as part of a sequence in developing improved diaphragm formulations prior to the transfer to plant cells.

5 PLANT EXPERIENCE

Laboratory testing in 1986 led to the first installations, in 1987, of experimental diaphragms in Columbia N-6 cells (1.5 kA/m^2: 30 kA, 20 m^2) at PPG's Natrium, West Virginia, plant. The decision was later made to fully convert this circuit to Tephram® diaphragms.

Figures 2 and 3 show, respectively, the individual voltage and current efficiency performance of each cell of the Columbia N-6 circuit as a function of time. Oxy “6” current efficiencies are reported. These have been corrected to neutral brine feed. The data are further normalized to 50% brine depletion and 30 kA operation. Four generations of Tephram diaphragms and a variety of doping strategies are represented in the composite data. The levels of hydrogen in chlorine are typically less than 0.1 vol.%.

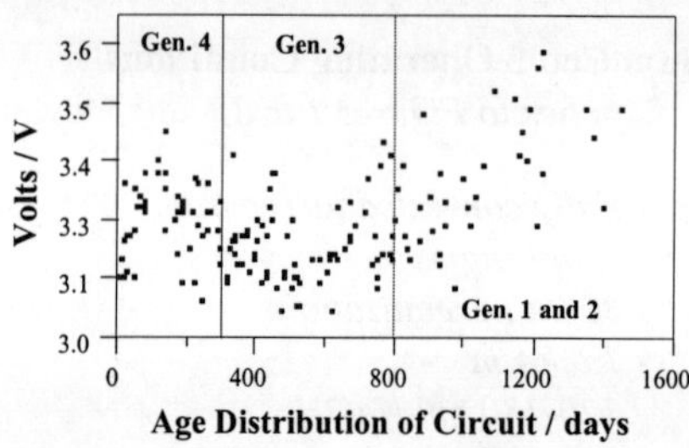

Figure 2 *The April 1997 voltage performance of the entire population of 160 Columbia N6 cells at Natrium Circuit 5 normalized to 30 kA. Each point indicates an individual cell. Voltages taken cell-to-cell.*

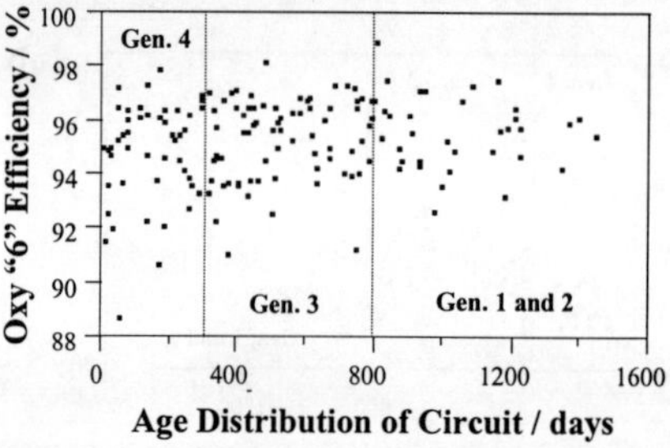

Figure 3 *The April 1997 current efficiency performance of the entire population of 160 Columbia N6 cells at Natrium Circuit 5. Each point indicates an individual cell. Efficiency is corrected to neutral brine feed.*

At this writing, some N-6 cell assemblies at Natrium are over 50 months old, see Figures 1 to 3. Cells have remained on line as long as 60 months. Mechanical aspects of these cells (such as gaskets, base mats, and anode stems) are becoming life limiting. In several cases, Tephram diaphragms have been reused after cell refurbishment. No degradation of the base diaphragm is seen.

Plant operations have shown that at the end of life, the washing-off of Tephram diaphragms with high pressure hoses is easier than for resin-modified asbestos diaphragms. The diaphragms tend to come off faster and in larger pieces, thus simplifying this operation.

PPG’s conclusions from Natrium’s N-6 trials are that Tephram diaphragms are resistant to brine upsets and outages. The option is available to brine flush during outages. With brine flushing, advantage can be taken of an outage to revise or restart a doping program at a reduced anolyte-to-catholyte level. In all, a circuit can be reliably run, with the flexibility to respond to unforeseen events.

Recent Generation 4 results from Natrium’s OxyTech MDC-55 trials are shown in Figures 4 to 10. Natrium’s MDC-55s are operated at a higher current density than the N-6s (2.2 kA/m^2: 120 kA, 55 m^2). The trials started in April, 1995; however, the higher current density conditions lead to the need to adapt a Tephram diaphragm formulation to the circuit. (References 10 and 11 discuss PPG's initial results with MDC-55s.)

Figures 4 to 10 show data from the first two cells in which the newly-introduced Generation 4 was installed. Generation 4 is PPG's evolutionary improvement which offers a better current efficiency in larger cell designs and certain start-up conveniences. Figure 4 depicts cell voltage vs. time. Figure 5 depicts current efficiency versus time. Oxy "6" current efficiencies are used. (Near neutral brine, pH 6 to 6.5, is now used at Natrium, which raises current efficiency by about 1% *vis-à-vis* alkaline brine. Previously published data [11, 12] date from when Natrium was using alkaline brine.) The data presented are normalized to 50% brine depletion and 120 kA operation. Cell liquor strength is shown in Figure 6. Oxygen in chlorine is shown in Figure 7 and hydrogen in chlorine is shown in Figure 8 (Natrium's detection limit is 0.05 vol.%). Chlorates in cell liquor are shown in Figure 9. The values shown are considered acceptable. Figure 10 tracks the "porosity factor" of the two diaphragms. PPG defines porosity factor as level (in inches) divided by flow (in gallons per minute) times 10. This is a useful way to track diaphragm permeability.

Natrium Cct. 8 Operating Conditions

(Applies to Figures 4 to 10)

- *Two Gen. 4 diaphragms*
- *120 kA normalization*
- *2.2 kA/m^2*

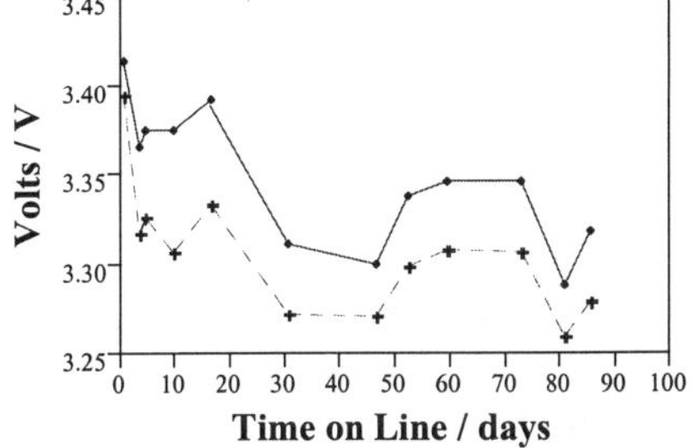

Figure 4 *The voltage performance of two Generation 4 Tephram® diaphragm cells in Natrium Circuit 8 normalized to 120 kA. Voltages taken cell-to-cell.*

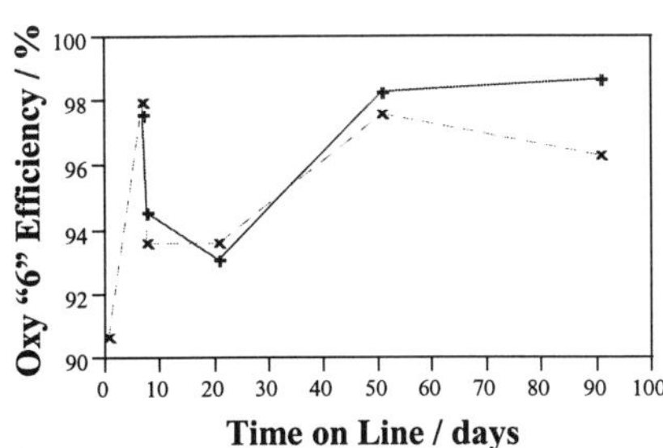

Figure 5 *The current efficiency performance of two Generation 4 Tephram® diaphragm cells in Natrium Circuit 8. Efficiency is corrected to neutral brine feed.*

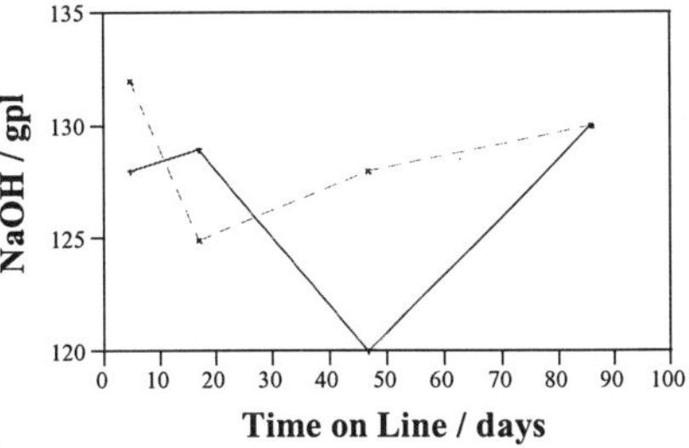

Figure 6 *The cell liquor strength for two Generation 4 Tephram® diaphragm cells in Natrium Circuit 8.*

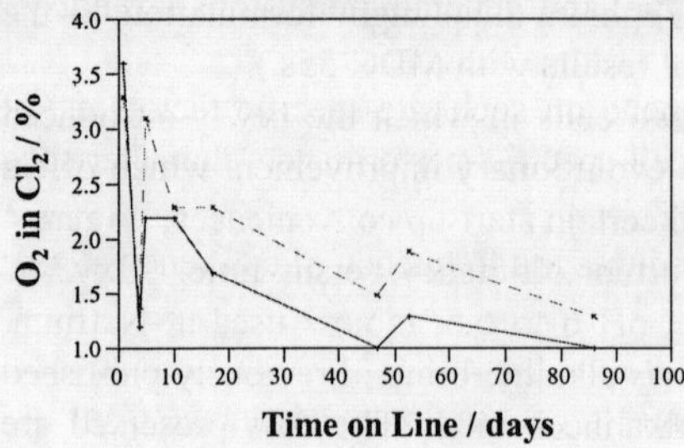

Figure 7 *The oxygen in chlorine for two Generation 4 Tephram® diaphragm cells in Natrium Circuit 8.*

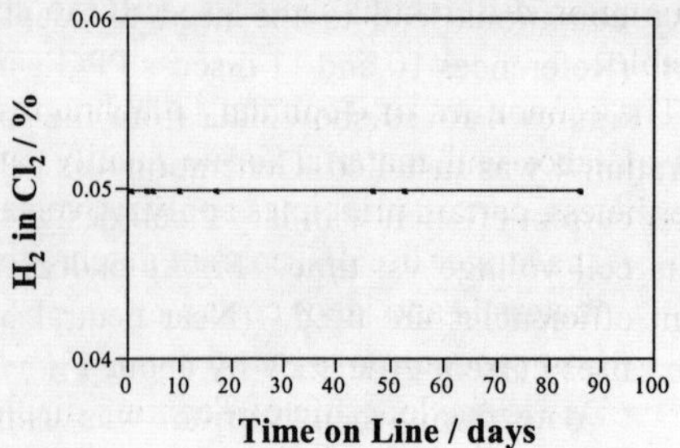

Figure 8 *The hydrogen in chlorine for two Generation 4 Tephram® diaphragm cells in Natrium Circuit 8.*

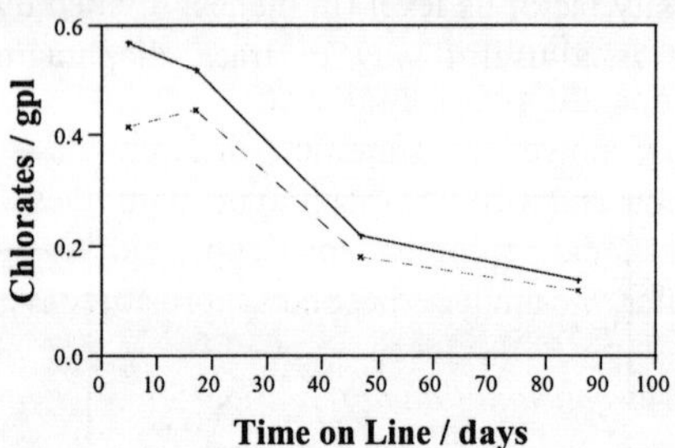

Figure 9 *The chlorates in cell liquor for two Generation 4 Tephram® diaphragm cells in Natrium Circuit 8.*

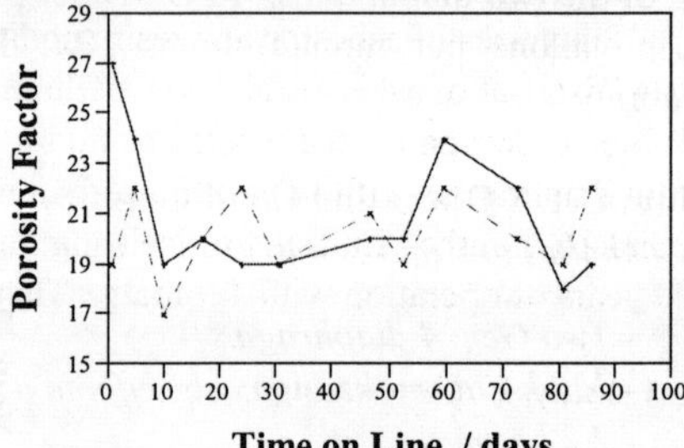

Figure 10 *The Porosity Factor for two Generation 4 Tephram® diaphragm cells in Natrium Circuit 8. PPG defines Porosity Factor as Level (in inches) divided by Flow (in gpm) x 10.*

Following the installation of Tephram® diaphragm deposition capability at PPG's Lake Charles plant, experimental trials in PPG's (Glanor) V-1244 electrolyzers were begun in February, 1996. Tephram diaphragms were installed on three elements of a 12-element bipolar electrolyzer. Initial results for these four foot tall diaphragms showed a 100 mV advantage in cell voltage, Oxy "6" current efficiencies of 95%, with essentially no hydrogen in chlorine. These results were extremely encouraging and have led to the installation of Tephram diaphragms in four full 12-cell electrolyzers. Two of these full electrolyzers use Generation 3 diaphragms and two use the newly-introduced Generation 4.

6 THE ECONOMICS OF TEPHRAM® DIAPHRAGM USE

Tephram diaphragms, being composed of man-made materials, are bound to have higher materials costs than asbestos diaphragms. Thus, a full and complete comparison cannot be made only on the basis of initial costs. The materials cost of Tephram diaphragms must be analyzed over the multi-year lifetime of the diaphragms and compared to the full life-cycle costs of asbestos, maintenance, and cell renewal labor. Any power consumption differential must also be factored in. Plant experience has shown that the power consumption differential can be specific to cell type, and general conclusions are not yet available.

The economics of Tephram® diaphragms depend on such site-specific factors as the costs of labor and material, brine quality, and cell-specific performance characteristics. Nevertheless, certain principles apply universally:

- The longer life of Tephram diaphragms reduces the frequency of cell renewal materials and labor costs.
- Lost production is reduced.
- Due to the long life of Tephram diaphragms, to get maximum advantage from them, all mechanical components of a cell must be in excellent condition prior to the installation of the diaphragms.
- Materials costs ($85-105/m^2$ of diaphragm) are higher than for asbestos and the cost of licensing the technology from PPG (a one time fee of $280/m^2$ of diaphragm) must be considered.
- Purchases of asbestos and resin modifiers are eliminated.
- Disposal of asbestos and water filtration operations are eliminated.

Table 1 compares the costs, excluding any power consumption differential, for Tephram diaphragms with lives of 36 and 48 months and asbestos diaphragms with lives of 8, 12, and 16 months. Entries for licensing, interest, and equipment costs apply only to the first 10 years of operation with Tephram. Thereafter, the analysis becomes more attractive.

Table 1 *Analysis of operating costs without a power consumption differential (\$/mT Cl_2)*

	Tephram® Life (Months)			SM-2™ Resin Modified Asbestos Life (Months)		
	36	48		8	12	16
Diaphragm Cost	1.62	1.21		1.99	1.33	1.00
Cell Renewal Cost	0.44	0.33		1.99	1.33	0.99
Licensing Cost	1.17	0.88		--	--	--
Equipment Cost	0.27	0.20		--	--	--
Interest Cost	0.78	0.58		--	--	--
Lost Production	0.05	0.04		0.21	0.14	0.10
Waste Disposal	--	--		0.22	0.15	0.11
Total	4.33	3.24		4.41	2.95	2.20

Assumptions: Paid-up license fee of $280/m^2$, license and equipment cost on basis of 10-year depreciation (8% interest), equipment cost of \$500,000 (depreciated similarly) for a 500 mT Cl_2/day, OxyTech MDC-55 plant operated at 120 kA and 95% current efficiency. Lost Production accounts for time out of service during cell renewal.

7 THE FUTURE

Tephram® diaphragms have now been tested in Columbia N-6s, Columbia N-3s, Hooker H-2As, Hooker H-3s, OxyTech MDC-29s, OxyTech MDC-55s, and PPG (Glanor) V-1244s. Operating experience in all these different cell designs is accumulating rapidly. PPG's Natrium Circuit 5 is continuing to set diaphragm cell lifetime records.

Tephram diaphragms respond to the desire of the chlor-alkali industry to find an alternative to asbestos that extends the interval between cell renewals and lowers the cost and risk of asbestos handling and disposal. Tephram diaphragms have enjoyed a favorable reception from potential licensees. (For the time being, PPG is depositing all diaphragms centrally at Natrium and Lake Charles.) The continued long-term evaluation of Tephram diaphragms by plants that are environmental leaders is forecasted. License agreements based on a one-time fee and access to on-going improvements are envisioned. There are five issued U.S. patents on Tephram diaphragms and other patents are pending. [5-10]

8 ACKNOWLEDGMENTS

The authors would like to acknowledge the participation of many PPG co-workers: at the Chemicals Group Technical Center, C. Dilmore (ret.), A. Maloney, J. Kinney, S. Pickens, and J. Snodgrass (ret.), K. Copeland, and J. Mauro; at Natrium, D. Bush, C. Blakely, J. Maxwell, C. Hill, and R. Toumala; at Lake Charles, J. Cimini, T. Jeffery (ret.), and S. Richardson.

9 REFERENCES

1. T.F. Florkiewicz and R.C. Matousek, *Polyramix™: A Non-Asbestos Diaphragm Separator*, The Chlorine Institute, 31st Plant Manager's Seminar, New Orleans, LA, March 9, 1988.
2. T.F. Florkiewicz and R.L. Romine, *Non-Asbestos Diaphragm Technology*, The Chlorine Institute, 35th Plant Manager's Seminar, New Orleans, LA, March 18, 1992.
3. F. Kuntzburger, D. Horbez, J.M. Perineau and J. Bachot, *New Developments in Microporous Built-in Precathode and Asbestos-Free Diaphragm*, "Modern Chlor-Alkali Technology", SCI, London, 1995, **6**, 140.
4. C.R. Dilmore and D.W. DuBois, *PPG's Tephram™ Synthetic Diaphragm Circuit Conversion*, "Modern Chlor-Alkali Technology", SCI, London, 1995, **6**, 133.
5. D.W. DuBois and W.W. Carlin, U.S. Pat. 4,720,334, 1988.
6. S.R. Pickens, D.W. DuBois and H.-C.M. Yang, U.S. Pat. 5,030,403, 1991.
7. C.R. Dilmore and B.A. Maloney, U.S. Pat. 5,188,712, 1993.
8. D.W. DuBois and C.R. Dilmore, U.S. Pat. 5,192,401, 1993.
9. D.W. DuBois, B.A. Maloney and S.R. Pickens, U.S. Pat. 5,567,298, 1996.
10. C.R. Dilmore and J.O. Snodgrass, PCT Int. Appl. WO 97 04,883, 1997.
11. P.C. Foller and D.W. DuBois, *PPG's Non-Asbestos Diaphragm for the Chlor-Alkali Industry*, The Chlorine Institute, Inc., 72nd Annual Meeting, Washington, D.C., March 27, 1996.
12. P.C. Foller and D.W. DuBois, *PPG Technology Journal*, 1995, **3(1)**, 49.

Tephram® is a registered trademark of PPG Industries, Inc.
Nafion® is a registered trademark of E.I. DuPont de Nemours and Co.

20
LONG LIFE DIAPHRAGM CELL

Thomas F. Florkiewicz

OxyTech Systems, Inc.
100 Seventh Ave.
Chardon, Ohio, U.S.A. 44024-1000

1 INTRODUCTION

The attention in chlor-alkali electrochemical cell technology has focused on membrane cells over the last decade. However, diaphragm cells continue to play a significant role in world wide chlorine production. For example, China and Russia continue to be dominated by diaphragm cell technology. Europe remains dependent on diaphragm cells with approximately 26% of European chlorine production carried out in diaphragm cells. In North America, diaphragm cells are the primary technology of use, accounting for roughly 70% of all United States of America (USA) production.

But why spend time and money to invest in diaphragm cell technology if membrane cells are the technology of choice? Three major reasons dictate the continued use of diaphragm cells:

1) Approximately 60% of large diaphragm cell plants in the USA depend on some form of co-generated electric power and steam. Many large European diaphragm cell plants also use some form of co-generated power and steam. It is known that steam from a co-generation plant normally has a very low cost. Because of the low strength caustic soda solution (cell liquor) produced from a diaphragm cell vs. the 32-35 weight % caustic soda from a membrane cell, the economics of conversion from diaphragm cell to membrane cell technology are very dependent on steam costs. Low steam costs make it extremely difficult to economically justify a conversion to membrane cell technology. Thus, unless fuel prices escalate in the future with a resultant increase in steam costs, or there is some additional benefit, it will be difficult to justify replacement of diaphragm cells with membrane cells.
2) Chlor-alkali companies continue to stress being the low cost producer. When a low cost producer wishes to increase capacity, the most economic method is by incremental expansion. The cost of a new (greenfield) chlor-alkali project is estimated to be 275,000 to 325,000 United States dollars (USD or $) per metric ton (MT) - day of chlorine capacity. An incremental expansion can be executed for an estimated cost of $200,000 to $250,000 per MT - day of chlorine capacity.

Thus, the low cost producer who has sufficient existing capacity would be wise to expand his facilities rather than undertake a greenfield project. Of course, the success of an incremental expansion depends on the condition of the whole diaphragm cell plant and not just the diaphragm cells.

3) Diaphragm cells can utilize inexpensive well brine as the salt source.

For these three reasons, OxyTech continues to invest in diaphragm cell technology. Improved diaphragm cell performance makes it more difficult to replace the technology with membrane cells and it makes incremental expansions that much more attractive to the producers.

To satisfy the needs of the diaphragm cell plant producers, OxyTech has developed a strategy to design a "Long Life Diaphragm Cell." In developing such a strategy, we must ask the question — has diaphragm cell technology reached an advanced level sufficient to support the continuous operation of a diaphragm cell for three to five years while operating at current densities of 2.5 - 3.0 kA/m^2? OxyTech believes the answer is yes!

Recent improvements have increased the life of major diaphragm cell components to achieve long life. There are, however, some components which require further improvement to achieve a cell life of three to five years. Of course, in undertaking a strategy to achieve a long life diaphragm cell, it must also be economically attractive to the producer. Thus, the purpose of this paper is to examine the present state of the technology to achieve a three to five year diaphragm cell, to state where additional development work is required, and to examine the economic benefits to the end-user.

2 CATHODE STRUCTURE

As diaphragm cell chlor-alkali producers push for more chlorine production to satisfy incremental increases in capacity, diaphragm cell current densities are being pushed higher and higher — now approaching 3.0 kA/m^2. Pushing more current through the cell results in additional heat generation through the conversion of electrical energy into heat, represented by the relationship I^2R. If this extra heat is not removed or redistributed, then the cell will operate at higher temperatures. In addition, diaphragm cell producers are moving towards higher and higher cell liquor concentrations to squeeze more capacity from the caustic soda evaporator. Cell liquor concentrations from 140-165 gpl are not uncommon. However, higher temperatures and higher caustic concentrations are not without a penalty. It is well known that steel will stress crack when exposed to high caustic concentration and temperature. Accordingly, OxyTech now fabricates MDC series diaphragm cells with extra copper and patented "current distributor bars". The bars are directly welded from the tubesheet to the steel sideplate to provide additional current carrying area as well as a more direct current path. The additional area and more direct current path reduces the temperature on the steel cathode flange and minimizes stress cracking. The benefit of current distributor bars is illustrated in Table 1. Installation of bars on a MDC-55 cell reduced the steel cathode flange temperature an average of 25°C on the front side steel sideplate above the copper gird bar, which is normally the hottest point on the cathode. The use of current distributor bars has been standard practice in the fabrication of the MDC series cells for the last ten years. They enable the cathode to achieve long, maintenance free performance at higher current

Table 1 *Temperature Readings on MDC-55 Cathode, With and Without Current Distributor Bars ("CDB")*

Measurement Location	Temperature Readings, °C	
	With "CDB"	Without "CDB"
TI Frontside Steel Sideplate Above Copper Gird Bar	91-93°C	110-125°C
T2 Frontside Steel Sideplate Below Copper Gird Bar	94-95°C	99-103°C
T3 Backside Steel Sideplate Above Copper Gird Bar	92-93°C	93-97°C
T4 Backside Steel Sideplate Below Copper Gird Bar	96-97°C	98-100°C

- Operating Current - 146 kA (2.65 kA/m^2)
- Using Hughes Probeye 686 Thermal Data Viewer

densities and higher cell liquor concentrations. Since current distributor bars are standard design practice for MDC series cells, there is no added cost to the end-user.

3 CELL HEADS

Years ago cell head construction relied on granite or concrete. Since the late 1960s, glass reinforced polyester (GRP) has been the material of choice. Other materials such as epoxy, titanium, PVC, CPVC, or vinylesters have been tried with limited acceptance. It is well known that GRP will eventually erode/corrode in service so that relining is necessary in usually 1-3 years. Typically 2-3 relines are possible before the cell head must be replaced with a new one.

Recently, a patented material by B. F. Goodrich known as Telene[§] RIM polymer has been shown to have remarkable erosion/corrosion properties. This unique, corrosion resistant material can be used to mold extremely large parts. OxyTech, working in conjunction with B. F. Goodrich, has developed Telene RIM polymer technology to mold diaphragm cell heads. Extensive testing has proven that Telene RIM polymer has remarkable erosion/corrosion properties and it can withstand the harsh environment when exposed to hot, wet Cl_2 gas. Table 2 presents physical property data of a new Telene RIM polymer cell head and one that has operated for 4½ years. Over a period of 4½ years, the physical properties have changed slightly and the measured loss in wall thickness was negligible.

[§] Telene™ is a trademark of B. F. Goodrich for dicyclopentadiene resin.

Table 2 *Telene RIM Polymer Cell Head Physical Properties*

NEW vs 4½ YEAR PERFORMANCE			
	New	4½-Year	%
Compressive Strength, kg/cm^2	600	675	13
Tensile Strength, kg/cm^2	470	526	12
Tensile Modulus, kg/cm^2	19,000	18,360	-3
% Elongation	4.0	4.1	2

Figure 1 shows the rapid growth rate of Telene RIM polymer cell heads over the last several years. Projections for 1997 predict total cumulative cell head sales to climb above 1500.

The first Telene RIM polymer cell heads commercially manufactured are approaching seven years life. The data supports, and it is predicted that, they will last for more than eight years. Thus, Telene RIM polymer cell heads achieve a lifetime which meets the criteria of a 3-5 year diaphragm cell.

From an economic standpoint, a Telene RIM polymer cell head will initially cost more to the end-user. The cost of GRP or other resin based cell heads vary with quality and fabrication. Table 3 compares the operating cost of a GRP cell head versus a Telene RIM polymer cell head. When economically compared to the GRP cell head, the Telene RIM polymer cell head, because it is maintenance free, is economically more attractive to the end-user.

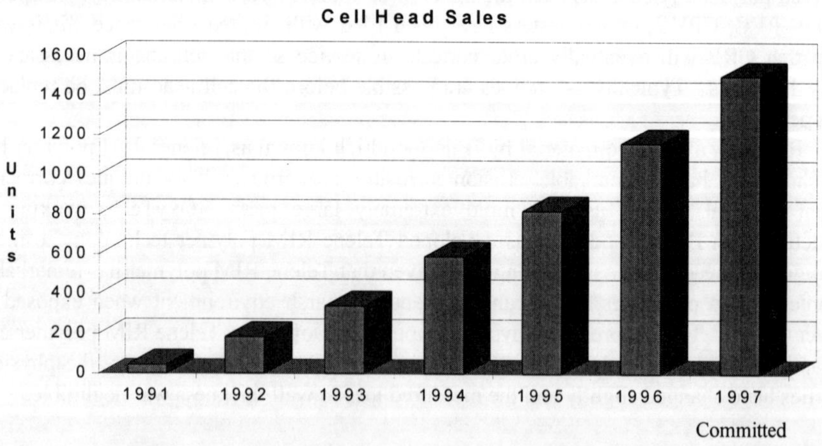

Figure 1 *Cumulative Telene RIM Polymer Cell Head Sales Growth*

Table 3 *Comparison of Operating Cost - GRP vs. Telene RIM Polymer Cell Head*

	GRP	Telene
Initial Cost	\$4,500	\$6,200
No. of Relines	Two	-
Cost of Reline	\$2,000	0
Life Before Replacement	6 Years	8 + Years
Operating Cost, \$/MT Cl_2	\$0.90	\$0.50

- MDC-55 Cell

One last benefit which should not be overlooked is the excellent resistance to corrosion/erosion of Telene RIM polymer material versus GRP. Have you ever thought of where all the polyester and glass fiber reinforcement goes as a GRP cell head erodes/corrodes?

4 DIAPHRAGM

Presently, the deterioration of the asbestos based diaphragms (for a variety of reasons), is the main cause for cell renewal. The recent advent of non-asbestos diaphragms has allowed the diaphragm cells to operate for far longer periods of time. OxyTech's contribution to non-asbestos diaphragm cell technology is called the Polyramix [§] (PMX) diaphragm. Because of the high cost of any non-asbestos diaphragm when compared to asbestos, it is a necessity, from an economic standpoint, that the non-asbestos diaphragm operate for a minimum of three (3) years. OxyTech is committed to see non-asbestos diaphragm cell technology succeed and we foresee that environmental pressures will eventually drive the industry toward the use of non-asbestos diaphragm technology. Some trends which point to this theory are:

- Legislation in France to ban the use of asbestos by the year 2002,
- European legislation to ban the use of asbestos by the year 2010,
- The State of Rio de Janeiro in Brazil has banned the production of caustic soda in cells using either mercury or asbestos, and
- Reduced number of companies mining asbestos grades suitable for diaphragm cell use.

The Vulcan Chemicals Plant at Geismar, Louisiana (600 MT/D Cl_2 capacity) has been operating with 100% Polyramix diaphragms for the last four (4) years. They are now achieving average diaphragm lives of greater than 1000 days, while operating at an average current density of 2.55 -2.65 kA/m^2. The reason for their success is three-fold:

[§] Polyramix® is a registered trademark of OxyTech Systems, Inc. for a polymer fiber composed of zirconia and Teflon resin.

1. The PMX diaphragm, composed of only zirconia and Teflon[§] resin is completely inert to the harsh atmosphere of both the anode and cathode side of the cell. Thus, there is no chemical attack on the PMX diaphragm. Since the diaphragm is fused prior to use, it is able to retain fiber strength much better than an asbestos diaphragm or a polymer modified asbestos diaphragm, or any other type diaphragm which is not fused. Table 4 presents tensile strength data showing the difference between a PMX diaphragm and an SM-2 or HAPP polymer modified asbestos (PMA) diaphragm. The higher integrity or strength of the PMX diaphragm prevents gas bubbles and anolyte from eroding the surface of the PMX diaphragm.

2. A PMX diaphragm, other non-asbestos diaphragms, and asbestos based diaphragms are all subject to pluggage of pores with brine contaminants. These contaminants are usually calcium (Ca), magnesium (Mg), iron (Fe), or nickel (Ni). OxyTech has developed patented procedures for essentially "washing" the diaphragm with inhibited hydrochloric acid (HCl). By adjusting the time and temperature of the inhibited HCl, any of the four contaminants can be removed from the PMX diaphragm. And, because PMX diaphragms are inert (containing only zirconia and Teflon resin), a flushing with inhibited HCl does little harm to the PMX diaphragm. Inhibited HCl is specified to retard any corrosion of the steel cathode screen.

3. Because of the high tensile strength of PMX diaphragms, procedures have been developed to efficiently remove a cathode with a PMX diaphragm from an old base and install it on a new base. So, even if the end-user must replace a base cover or base for any reason (failed gasket, anode stud leak, old base cover), it can be done without injury to the PMX diaphragm. Success rate for these procedures is now approximately 90%.

Table 4 *Tensile Strength of Diaphragms*

Plant	Diaphragm Type	Tensile Strength, kg/cm^2
A	SM-2	1.70
B	SM-2	3.25
C	SM-2	3.13
D	HAPP	1.16
Various	PMX	10.0 (Average)

- Using an Instrom Tensile Tester Unit, Model 4411

§ Teflon® is a registered trademark of E. I. DuPont de Nemours and Company for flurocarbon resins.

It is a fact that non-asbestos diaphragms cost substantially more than PMA diaphragms. The difference could be up to twenty (20) times the cost of a PMA diaphragm. To offset this difference a three year life for a non-asbestos diaphragm may be considered minimal, especially if it cannot demonstrate other operating savings such as lower voltage. A typical analysis of OxyTech's Polyramix diaphragm versus a PMA diaphragm is shown in Table 5.

By extending the life of Polyramix diaphragms to five years, the end-user can gain an operating cost saving of $1.60 MT Cl_2 even if there are no voltage savings or no additional costs associated with asbestos handling and disposal.

Thus, long life for a non-asbestos diaphragm has been commercially demonstrated. A three-year life has been achieved at the Vulcan-Geismar, Louisiana chlorine plant. Demonstration blocks of 10 to 20 PMX diaphragm cells in Europe have achieved over five (5) years life and the oldest PMX diaphragm cell at the Occidental Chemical Deer Park, Texas chlorine plant is now operating for over eight (8) years.

5 GASKETS

Gaskets may be separated into two distinct groups — base covers and perimeter gaskets. Perimeter gaskets are defined as the gaskets which seal the cell head to the cathode and the cathode to the base. With long lives being achieved by non-asbestos diaphragms, it becomes evident that the highest cause of failure of the diaphragm cell is the gaskets. And it is these gaskets which now pose the biggest challenge to overcome if a 3-5 year cell life is to be achieved.

The base cover gasket is traditionally made of rubber — usually EPDM (ethylene propylene diamine) rubber. The life of the base cover is usually one to three diaphragm changes or about 600 days, and it costs approximately $600 for an MDC-55 cell. Many types of thin sheet titanium base covers have been tried. OxyTech's efforts in using a thin sheet of titanium is called a Tibac[§] base cover. The Tibac base cover must still depend on

Table 5 *Comparison of Operating Cost - PMA vs. PMX Diaphragm*

	PMA Diaphragm	PMX Diaphragm	
Life, years	0.75	3 Year	5 Year
Operating Cost, $/MT Cl_2			
- No Voltage Savings	$4.13	$4.21	$2.53
- 0.1 Volts Savings Over PMA Diaphragm	$4.13	$2.11	$0.43

- MDC-55 cell $@ 2.55 kA/m^2
- Power Cost - $0.03/KWH

[§] Tibac is a trademark of OxyTech Systems, Inc. for a titanium base cover for diaphragm cells.

rubber seals around the perimeter and on rubber seals at the anode collars. To further improve the titanium base cover, OxyTech's latest thrust is to develop what are termed welded base covers. As shown in Figure 2, the titanium anode collar is welded to a titanium sheet to provide a gasket-free assembly.

Such a design must allow sufficient flex of the anode during assembly so that the base cover/anode weldment will last as long as the anode coating. Since anode coating life of 10-15 years is commonplace in diaphragm cells, the combination of the welded titanium base cover and titanium anode would only require replacement when the anodes are recoated. The only area now dependent on a rubber component would be the perimeter gaskets of the cell head to cathode and the cathode to base.

The design of these two perimeter gaskets is a very difficult challenge facing diaphragm cell technology suppliers to reach a three to five year cell life. One of the major concerns, apart from the corrosion of a gasket, is the ability to seal when the cell temperature cycles, due to an increase/decrease in current load or an outage (whether planned or unplanned). This source of leakage causes more corrosion to the cathode and to the external copper than any other mechanism. OxyTech's effort in this area is now concentrating on the design of a pneumatic gasket similar to that depicted in Figure 3. It is an "inflatable" gasket, similar to a bicycle tire. It is of two-piece construction - the smaller inflatable gasket and the larger sealing gasket. If the gasketing pressure changes due to gasket compression set from changes in cell temperature, air can be added to the inflatable gasket to restore the compressive sealing force on the main sealing gasket. Presently, the pneumatic gasket is in the prototype stage.

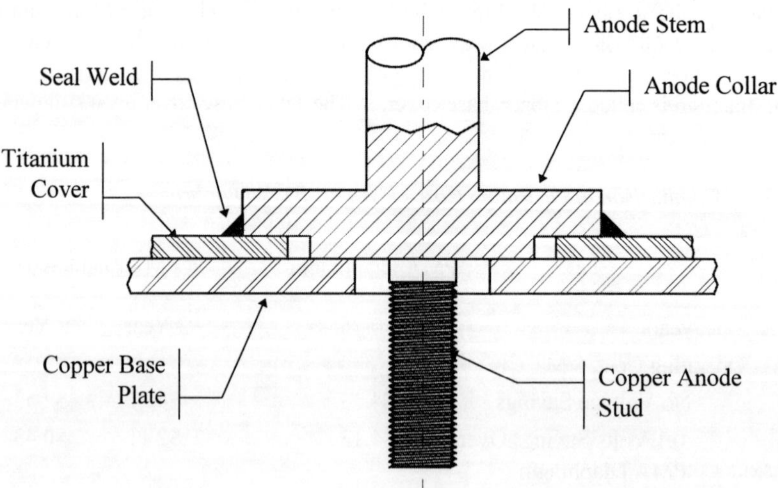

Figure 2 Welded Titanium Base Cover

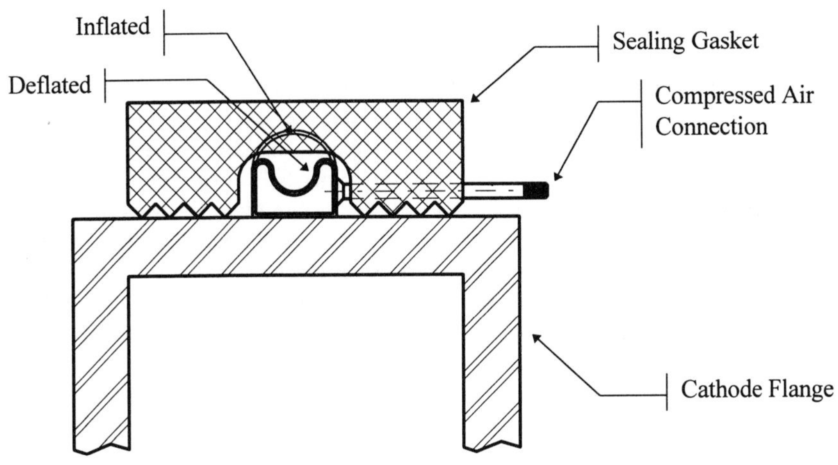

Figure 3 *Pneumatic Gasket Design*

Our preliminary work indicates that a pneumatic gasket system will be rather expensive, perhaps a factor of ten times the cost of a straight EPDM gasket. However, the pneumatic gasket portion of the system should be re-usable. The justification will have to be borne out over several years of leak-free service. The operating cost of an EPDM base cover plus perimeter gaskets is compared to a Tibac base cover and pneumatic gaskets in Table 6.

Table 6 *Comparison of Operating Cost — EPDM Gaskets vs. Tibac Base Cover & Pneumatic Gasket*

	EPDM Gaskets	Tibac Base Cover & Pneumatic Gasket	
Life	600 Days	3 Year	5 Year
Base Cover Cost, $	$600	$2,000	$2,000
Perimeter Gasket Cost, $	$300	$2,000	$2,000
Labor Cost, $	$600	$ 800	$ 800
Operating Cost, $/MT Cl_2			
Base Cover	$0.50	$0.60	$0.36
Perimeter Gasket	$0.12	$0.88	$0.53
Total	$0.62	$1.48	$0.89

- MDC-55
- Current Density - 2.55 kA/m^2

6 BENEFITS OF LONG LIFE DIAPHRAGM CELLS

Where do all these advances in diaphragm cell technology leave the end-user? Certainly, to achieve some of the advancements discussed, there is a higher cost versus the present level of technology (e.g. non-asbestos diaphragms vs. polymer modified asbestos diaphragms, standard vs. pneumatic type gaskets). But, is there a financial advantage to the end-user by choosing technology to allow the diaphragm cell to operate three to five years? Table 7 compares the cost of the advances discussed in this paper and compares it to a diaphragm cell using what is considered standard technology of the 1990s.

In the analysis, there are additional benefits derived by using advanced diaphragm cell technology. These benefits include a calculated saving of \$0.40 per metric ton of chlorine by not incurring handling and disposal costs of asbestos. Also, a long life diaphragm cell reduces the number of cell renewal operators required, requires less use of jumper switches, and also results in greater ECU production by reducing the time a renewed cell is jumpered out of service. These savings are conservatively estimated at \$0.25 per metric ton of chlorine.

The conclusion is that a three year diaphragm cell is marginally beneficial to the end-user. By extending the life to five years there is a substantial economic benefit to the end-user. Advanced technology is now available to achieve a three or five year life for the cathode, cell head, and diaphragm. Continued work on cell base covers and gaskets should complete the remaining parts of the puzzle and give the end-user the advanced designs necessary to reach the next plateau in diaphragm cell technology, which is a three to five year diaphragm cell life.

Table 7 *Comparison of Operating Cost — Standard vs. Advanced Diaphragm Cell Technology*

	OPERATING COST, \$/MT Cl_2		
	Standard Technology	Advanced Technology	
		3 Year Life	5 Year Life
Cathode Structure	\$2.10	\$2.10	\$2.10
Cell Head	0.90	0.50	0.50
Diaphragm	4.13	4.21	2.53
Gaskets	0.62	1.48	0.89
Lost Production	0.25	0.05	0.03
Asbestos Handling & Disposal	0.40	0	0
	\$8.40	\$8.34	\$6.05

21
NEW DEVELOPMENTS IN BUILT-IN PRECATHODE DIAPHRAGM TECHNOLOGY

F.Kuntzburger*, D.Horbez*, J.G. Le Helloco* and J.M. Perineau**

* Rhône Poulenc Recherches
52 rue de la Haie Coq, 93308 Aubervilliers (France)

** Rhône Poulenc Chimie
BP17, 38800 Le Pont de Claix (France)

1 INTRODUCTION

Over the past 30 years, the development of dimensionally stable anodes (DSA) and polymer modified diaphragm have contributed to a significant reduction in the energy requirement for diaphragm cell. Potential breakthrough of the technology will rely on significant improvements of the cathodic element including diaphragm structure optimization and cathodic activation. The patent literature covers many different types of active coatings which did not come into commercial application due to complicated manufacturing processes, expensive cost and poor life-time.

Rhône Poulenc Chimie (RPC) presented the "Built-in Precathode Diaphram" at the London Chlorine Symposium in June 1994. That concept opened both new way of cathode activation and diaphragm performance improvements.

This paper presents the concept, the main advantages, the industrial experience and the application field of built in precathode.

2 BUILT-IN PRECATHODE AND OPERATING PRINCIPLE

Precathode technology is based on the incorporation of a volumic cathode area in the first section of the diaphragm itself (Figure 1).

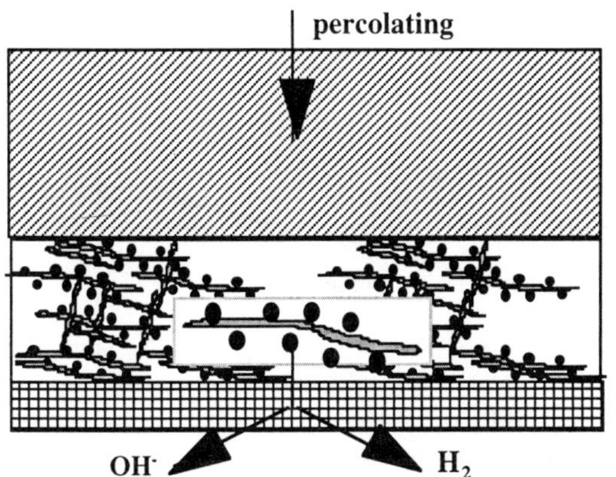

Figure 1 *Schematic of built in precathode diaphragm*

The precathode includes conducting fibers, electrocatalytic material, pore forming agents and fluoro-containing polymer binder.

Precathode in combination with diaphragm constructs a complete separator. The two layers are consecutively vacuum deposited onto a cell cathode in the same manner and with the same equipment as diaphragms.

The steel screen structure of the cathode box simply ensures the electricity conveying to the conductive fibres.

The electrocatalytic agent homogeneously dispersed in the volumic electrode is connected to the cathodic grid by conductive fibres. The electrochemical reduction of water takes place directly on the large active surface area of the electrocatalyst with lower overpotential.

3 CATHODIC ACTIVATION

3.1 Raney Nickel as Electrocatalyst

The electrocatalyst agent is incorporated into the precathode in the form of a powder whose particle size may vary from 1 to 100 µm. The catalyst is chosen from among the metals of the platinum group, Raney alloys, Raney metals and especially Raney nickel.

Due to its low cost and chemical stability in alkaline medium, nickel is widely used as cathodic material in the membrane cell.

Raney nickel type electrodes exhibit high catalytic activity for the hydrogen evolution reaction in alkaline media. Catalytic effect comes mainly from the high surface area formed after the leaching of aluminium from initial Ni-Al alloy leaving a two-scale porosity with nanopores and micropores.

Overpotential is dependant on electrode current density: it decreases when the surface area is increased. The electrochemically active surface of Raney nickel from Ni-Al alloy has been investigated to make a distinction between electrocatalytic and surface roughness effects[1].

The pore surface actually flooded by the electrolyte has been estimated by A.C. impedance. Voltammetric charge measurements on $Ni(OH)_2$/NiOOH redox system[1,2] have been used to evaluate the blocking effect of the evolved gas.

The Raney nickel particle exhibits a pore surface electrochemically active of 8 m^2/g. Cathodic polarization curves have been carried out on smooth nickel and pressed pellets of Raney nickel powder (Figure 2).

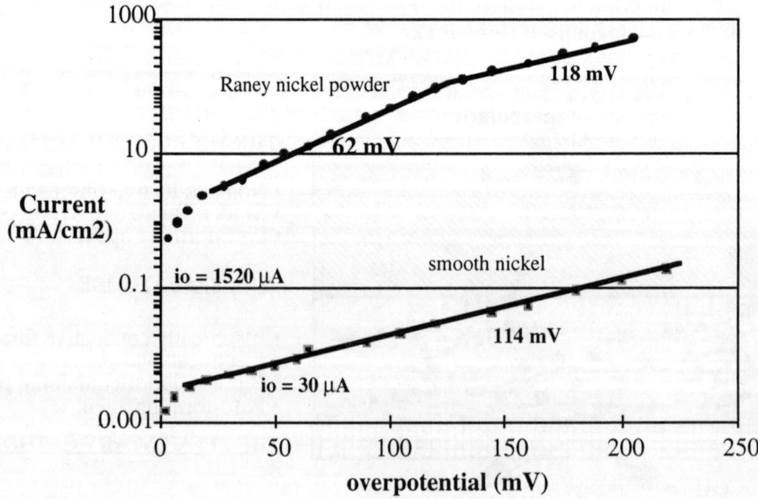

Figure 2 *Cathodic polarization curves on smooth Ni and pressed pellets of Raney nickel*

For smooth nickel the curve is linear in the whole potential scale with Tafel slope of 114 mV/decade of current and current exchange density of 30 μA/cm^2 (with a roughness factor of 3.8).

On Raney nickel powder the curves are bent with Tafel slopes of 60 mV/decade at low current density and a higher slope of 120 mV/decade at high current density. Those results are in agreement with the literature[3-5].

Differents explanations have been suggested such as an effect of very high hydrogene pressure in the nanopores (80 MPa) [6,7] and electrocatalytic effect involving chemical and electrochemical recombination mechanisms [4,8].

In this case, we can conclude that the nanopores of Raney nickel particles are partly electrochemically active because of ohmic or diffusion limitation, causing kinetics with a high apparent Tafel slope at high current density.

3.2 Precathode as a Volumic Electrode

The increase of electrochemically active surface is made possible by dispersion of the catalyst in a volumic electrode as the precathode.

The distribution of the electrocatalyst in the precathode is very homogeneous as is shown by the scanning electron micropscopy view (Figure 3).

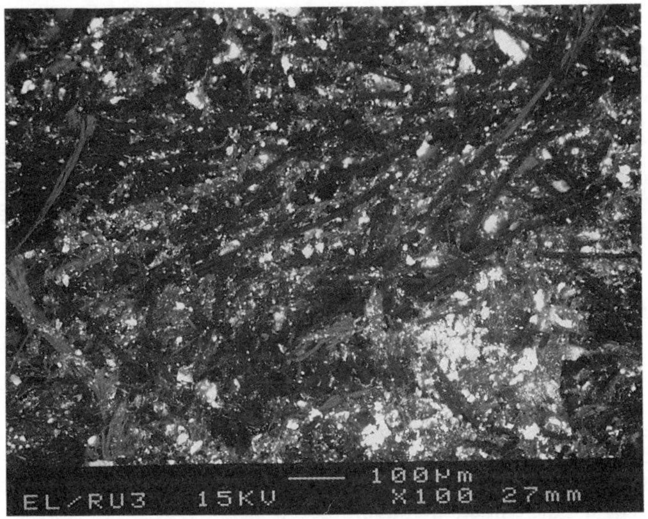

Figure 3 *Scanning Electron Micropscopy view of Raney nickel dispersion*

The precathode can perform almost as a plane electrode with an equivalent electroactive surface area. The effective current density is lowered due to homogeneous dispersion of the high surface catalyst. At low effective current density, ohmic drop has a weak influence on the Tafel slope as described by flow-through electrode model[9].

3.3 Stability of the Electrocatalyst after precathode Sintering

The precathode is heated to a temperature above the melting point or softening point of the fluoro-containing polymer used as the binding agent. So, it has to be controlled how the catalytic properties of Raney nickel are modified by the heating treatment.

The cathode potential for the hydrogen evolution on PTFE-based compact pellets, prepared without heating and with heating under air and argon are presented on Figure 4.

The catalytic properties of Raney nickel for the hydrogen evolution reaction are modified in a reversible way when the heating treatment is operated in inert medium.

In presence of oxygen, x-ray diffraction pattern shows NiO as oxidation product. NiO is reduced under cathodic polarization probably due to the high hydrogen pressure in the precathode. After 3 hours under hydrogen evolution, electrocatalytic properties are recovered with a loss of less than 40 mV at 300 mA/cm^2 (Figure 4).

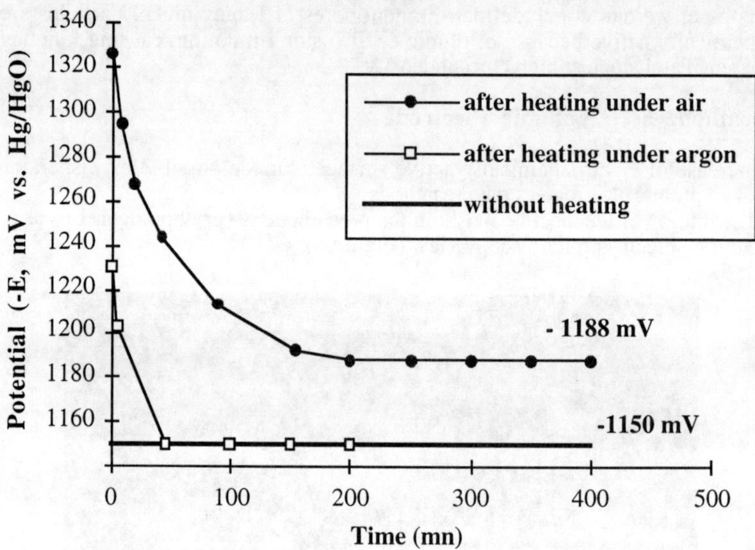

Figure 4 *Cathode potential for the hydrogen evolution on Raney Ni after heating treatment*

4 ELIMINATION OF CHLORATE ON PRECATHODE

The presence of chlorate in NaOH from diaphragm cells poses severe corrosion problems and also affects the product quality of caustic.

Chlorate is generated[10]

- at pH < 4, by anodic oxidation of hypochlorous acid

(kinetics depends on pH, hydrodynamics, anode coating)

$$6\ ClO^- + 3\ H_2O \rightarrow 3/2\ O_2 + 2\ ClO_3^- + 4\ Cl^- + 6\ H^+ + 6\ e^- \quad (1)$$

- at pH > 4, by chemical chloration in presence of hypochlorous acid

$$2\ HClO + ClO^- \rightarrow ClO_3^- + 2\ Cl^- + 2\ H^+ \quad (2)$$

The nature of the chlorate reduction reaction (3) proceeding on the cathode side is probably diffusion-controled due to the small concentration (0.05-0.02 M).

$$ClO_3^- + 3\ H_2O + 6\ e^- \rightarrow Cl^- + 6\ OH^- \quad (3)$$

Chlorate reduction on Fe-based material in an electrolyte simulating the catholyte in a diaphragm type chlor-alkali cell has been observed by Tilak *et al.*[11]. It is shown that chlorate is indeed reduced to chloride, the reduction being specific to only "Fe" and not metals such as Co, Ni, Mo, Ti, Hg, C in smooth form. It is feasible to enhance the rate of chlorate reduction using high surface-area Fe-based.

The elimination of chlorate on the cathode was determined during electrolysis under galvanostatic control (2.5 kA/m^2, 85°C). Different chlorate contents in anolyte were obtained by changing the catholyte composition (NaOH 2.5 - 4 M). Chlorate elimination is characterized by the difference between chlorate content in anolyte and catholyte.

Raney nickel content in precathode was taken as a parameter (100% is high content level).

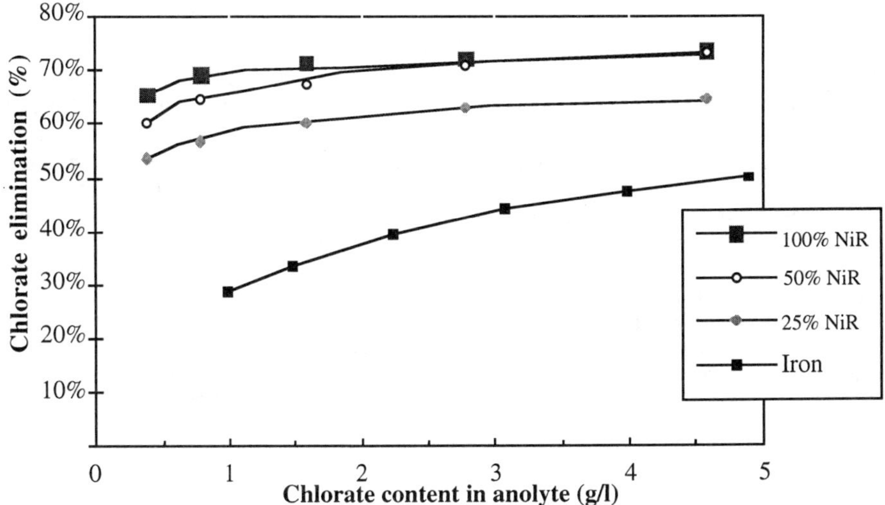

Figure 5 *Elimination of chlorate on precathode*

Performances of the precathode compared with non coated iron cathode are presented on Figure 5. Laboratory scale experiments on pilot cells showed that chlorate is reduced more effectively on precathode than on classical iron cathode. Chlorate from anolyte is cut by 60 to 75% after percolating through the precathode. This strong reduction of chlorate corresponds to decreasing of 50% in caustic soda.

High surface area of Raney nickel and high hydrogen pressure in the pores of the precathode enhance the elimination of the chlorate in the diaphragm cell.

5 ACTIVATION AND CATHODIC PROTECTION

The thermodynamic properties of metals in aqueous systems are commonly presented as Pourbaix diagrams in which equilibrium potentials for various reactions are plotted as a function of pH at the temperature of interest.

The published diagrams have been restricted to metal-water systems at room temperature and low concentration of Cl^-. More recently, theoretical relationships have been established which enable to construct Pourbaix diagrams for the Fe-Cl^--H_2O systems at 25 to 150°C[12]. The Pourbaix diagrams at 25, 60 and 100°C in high pH media are represented in Figure 6.

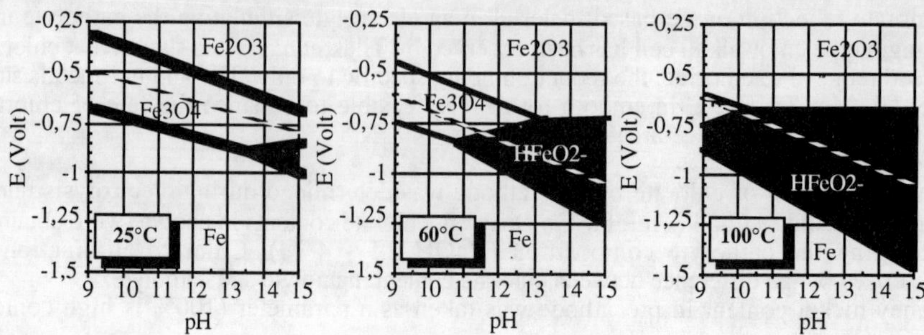

Figure 6 *Pourbaix diagrams for the Fe-Cl⁻-H_2O sytems at 25, 60 and 150 °C*

In comparison to the conventional Pourbaix diagram, the most noticeable change in the higher temperature diagrams is an expanded region of corrosion resulting from the increasing stability of dihypoferrite ion, $HFeO_2^-$. It has been suggested[13] that the effect of temperature on the appearance of potential-pH diagrams is generally slight, but the increasing relative size of the domain of the complex anion $HFeO_2^-$, is a notable exception[14]. In fact, increasing stability of complex ions in aqueous solutions at higher temperatures is not unusual and has been discussed[15] in terms of the decreasing dielectric constant of water.

The lowering of the hydrogen overvoltage is equivalent to an increase of the equilibrium concentration of $HFeO_2^-$ ion and corrosion of carbon steel cathode support.

Pourbaix diagrams indicate that preferential operating conditions to limit corrosion under cathodic activation are relatively high current density and moderate temperature.

Non oxidisable material such as stainless steel cathode mesh can also be used instead of carbon steel to operate at lower hydrogen potential in a larger temperature-current density field.

Cells are operating with stainless steel cathode mesh at RPC's Pont de Claix plant.

6 INDUSTRIAL APPLICATION OF PRECATHODE

6.1 RP Chimie Pont de Claix Plant

Built-in precathode diaphragm was first installed at RPC's Pont de Claix plant in 1986. The plant was fully converted in 1987.

Name plate capacity of this plant is 245 ktCl_2/year using only diaphragm cell (Hooker-S3B and Hooker-H4). Cells are equipped with modified diaphragm and expandable DSA® anodes. The brine supply is alkaline and not saturated.

Table 1 presents average performance data before and after the conversion to precathode technology.

6.2 HERAEUS experiment on MDC-55 cell

HERAEUS has installed the precathode technology within its own test facilities in MDC-55 cells (2.64 kA/m^2).

Table 2 shows the operating parameters in MDC-55 cells, with and without the precathode technology.

		With Precathode (average)	Without Precathode (average)
Current density	kA/m^2	1.7-1.9	1.7-1.9
Catholyte temperature	°C	75-80	75-80
NaOH concentration	g/l	130	130
Cell Voltage (without connection)	V	3.10 -3.20	3.25 - 3.35
Chlorate in caustic soda	g/l	0.2 - 0.3	0.5 - 0.6
Current efficiency	%	94 - 95	91.5 - 93.5
O_2 in chlorine	%	1.5 - 2.0	1.5 - 2.5
H_2 in chlorine	%	< 0.1	< 0.5
Power consumption	kWh/tCl_2	2465 - 2570	2625 - 2765
Average Energy saving	kWh/tCl_2	**175**	

Table 1 *Precathode Technology at RP Chimie Pont de Claix Plant*

		With Precathode (test average)	Without Precathode (test average)
Current density	kA/m^2	2.64	2.64
Catholyte temperature	°C	95	95
NaOH concentration	g/l	130	130
Cell Voltage	V	3.59	3.73
Chlorate in caustic soda	g/l	< 0.01	0.25
Current efficiency	%	98	95
O_2 in chlorine	%	0.95	1.8
H_2 in chlorine	%	< 0.1	0.2
Power consumption	kWh/tCl_2	2765	2965
Average Energy saving	kWh/tCl_2	**200**	

Table 2 *Technical Parameters of Precathode in MDC-55 (HERAEUS Test)*

6.3 Industrial Advantages of Precathode

According to the industrial experience in Rhône Poulenc and test results in MDC-55 in Heraeus Elektrochemie the advantages of precathode are following:

- cell voltage is reduced by 100-150 mV, with respect to current density from 1.7-2.6 kA/m^2
- oxygen in chlorine is decreased and current efficiency is increased by 0.7-3.0% due to the improvement of diaphragm and additional resistance of back migration of OH^- from catholyte to anolyte by precathode layer

• chlorate is cut by 30-50% in caustic soda due to strong reduction of ClO_3^- on precathode
• hydrogen in chlorine is lowered below 0.1% during standard operation
• safety is improved particularly during start up and shut down
• diaphragm life time is extended
• permeability of diaphragm is improved making high current density under constant anolyte level possible.

7 ECONOMICS

Energy saving by precathode is based not only on lowering cell voltage but also on increasing current efficiency. Both effect lower power consumption in diaphragm cell. According to the results in MDC-55 cell (Table 2), the power consumption saving by precathode is 200 kWh/tCl_2 at 2.64 kA/m^2.

The evaluation of cash value of yearly energy saving is shown graphically for different cathode productivities in Figure 7.

The savings are the pure energy saving - the extra benefits are not included.

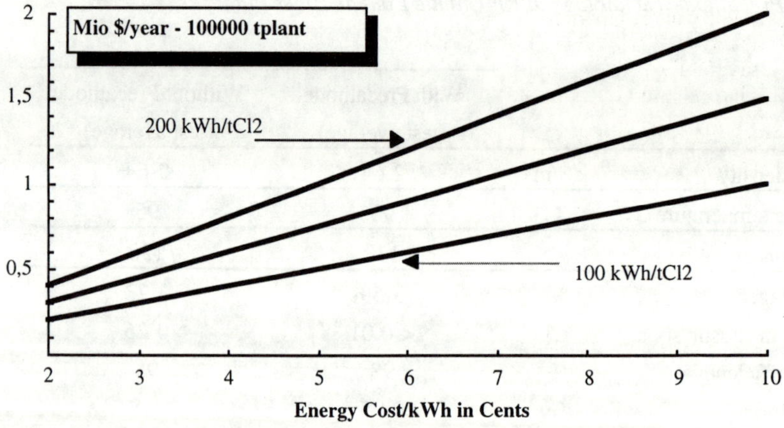

Figure 7 *Cash value of yearly pure energy saving with precathode*

On the other hand, the costs of precathode are far less than the resulting energy saving. The introduction of technology is very simple. No major additional equipment is needed. Thc life time of the diaphragm has been found to be improved by introduction of the precathode. The maintenance costs, therefore are reduced.

8 CONCLUSION

The original concept of a built-in precathode diaphragm contributes towards a reduction of electric power consumption of the diaphragm cell by cathodic electroactivation and improvement of current efficiency. The quality of anodic and cathodic products is improved by reduction of hydrogen content in chlorine and chlorate in caustic soda.

The technology keeps the diaphragm process attractive to today's chlor-alkali industry by significant advantages with minimum additional capital investment.

Rhône-Poulenc Chimie has come to an agreement with HERAEUS ELEKTROCHEMIE and DE NORA S.p.A for the world wide licensing of the precathode technology.

REFERENCES

1. J.G. Le Helloco, "Hydrogen evolution reaction on electrocatalytic powders in alkaline media", Thesis I.N.P.Grenoble, 1994.
2. D.M. Tench and E. Yeager, *J. Electrochem. Soc.*, 1973, **120.**
3. Y. Choquette, L. Brossard, A. Lasia and H. Menard, *J. Electroanal. Chem.*, 1990, **294**, 123.
4. Y. Choquette, L. Brossard, A. Lasia and H. Menard, *J. Electrochem. Soc.,* 1990, **137**.
5. Y. Choquette, L. Brossard, A. Lasia, and H. Menard, *Electrochem. Acta,* 1990, **35**, 1251.
6. S. Rauch and H. Wendt, *J. Applied Electrochem.*, 1992, **22**, 1025.
7. T. Burucinski, S. Rauch and H. Wendt, *J. Applied Electrochem.*, 1992, **22**, 1031.
8. M. Okido, J.K. Depo and G.A. Capuano, *J. Electrochem. Soc.,* 1993, **139**, 127.
9. A. Storck and F. Cœuret, "Elements de Génie Electrochimique", Ed. Lavoisier, Paris, 1984.
10. N. Ibl and H. Vogt, Comprehensive Treatise of Electrochemistry, Plenum Press, New York, 198, Vol. 2, Chapter 3.
11. B.V. Tilak, K. Tari and C.L. Hoover, *J. Electrochem. Soc.*, 1988, **135**, 1386.
12. S. Kesavan, T.A. Mozhi and B.E. Wilde, *Corrosion Science*, 1989, **45(3)**, 213.
13. R.M. Garrels and C.C. Christ, Solutions, Minerals and Equilibria, Harper and Row, New York, 1965.
14. H.E. Towsend, *Corrosion Science*, 1970, **10**, 33.
15. H.C. Helgeson, *J. Phys. Chem.*, 1967, **71**, 3121.

THE EFFECTS OF AQUEOUS DISPERSIONS ON CATHODE PERFORMANCE IN MERCURY CELLS

S. S. Povall

Technical Operating Plant Manager
JKL Units
ICI Castner Kellner Works
Runcorn, Cheshire

1 BASEPLATE THICK MERCURY - PROBLEM AND EFFECTS

One of the most significant causes of mercury cell efficiency problems is the formation of baseplate thick mercury, the extent depending on the rate of formation. At Castner Kellner Works, high rates of thick mercury formation are observed at current densities of >7 kA/m^2. Due to the effects of this, the manually adjusted cells are operated at increased inter electrode gaps and suffer the consequential increase in power consumption. At the end of the life of each cell the current distribution deteriorates and the mercury flow reduces. The combination of these end of cell life effects causes an increase in the cell gas hydrogen levels, cathodic (and anodic) current efficiency losses, and increases in anode coating wear rates. Cell treatments to decompose thick mercury cause higher cell loads to compensate for cell down times, and higher k factors when cells are restarted. Cell mercury losses also occur during the procedures to decompose thick mercury.

Baseplate film thickness measurements are measured directly using a dial gauge and a probe, together with a voltmeter connected to the probe and baseplate. The probe, insulated above the tip, assumes the potential of the brine (~2 volts). This reduces to zero when the tip of the probe contacts the mercury surface. The difference between the dial gauge reading at this position and the reading when contact with the baseplate is made gives the mercury film thickness. Changes in mercury film thickness may also be assessed from anode/cathode k factor changes that occur during the life of the cell.

Typical thick mercury formation rates at the eight anode cover positions in the J Unit cells are shown in Figure 1.

Figure 1 shows the marked load effect and the tendency for the position of the peak formation rate to move towards the mercury inlet end of the cell.

2 IMPURITIES IN BASEPLATE THICK MERCURY

Analysis has shown the most significant impurities found in baseplate thick mercury are sodium, strontium, and iron. Copper and molybdenum are also consistently found but an order of magnitude less in concentration. Analysis has also shown that little of the alkalinity is due to sodium amalgam.

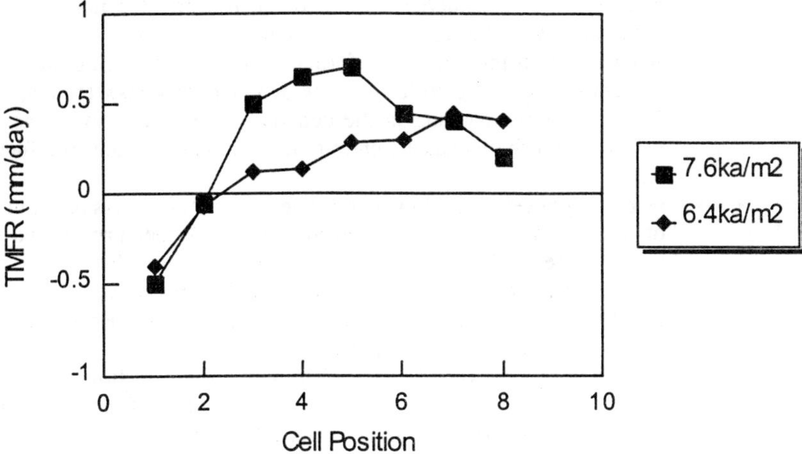

Figure 1 *Thick Mercury Formation Rates (mm/day)*

When thick mercury samples are taken carefully, from beneath the mercury surface, X-ray diffraction has shown the presence of sodium hydroxide in the anhydrous and monohydrate forms. The presence of anhydrous strontium hydroxide, strontium ferric hydroxide, $Sr_3Fe_2(OH)_{12}$, and mercuric oxide have also been shown. Gas, assumed to be hydrogen, is present too. These components comprise approximately 5% of the total volume.

3 FORMATION MECHANISM AND SUPPORTING EVIDENCE

The believed cause of baseplate thick mercury is the reaction of aqueous dispersions in the amalgam entering the brine cells with certain impurities that lead to almost dry crystalline particles. This dehydration of the droplets is probably assisted by fine oxide dispersions also present in the amalgam. The partially amalgamated particles dispersed in the amalgam agglomerate to form thick mercury which adheres to the cell baseplate.

This postulated mechanism is the result of the following observations:

- Analysis of a large number of thick mercury samples has shown that strontium compounds form a large proportion of the crystalline components of thick mercury particularly at the amalgam outlet end of the cell where most of the thick mercury is deposited. Formation of these strontium compounds is considered to be one of the principle mechanisms by which the dispersed aqueous droplets dehydrate. This was given credence when the strontium in the brine feed was reduced from approximately 7 ppm to 2 ppm. This caused the rate of thick mercury formation to reduce by at least 30% of that normally observed. Also, the thick mercury was confined to the outlet end of the cell.

- In the laboratory shaking dry, weak amalgam with a trace of air and the compounds found in thick mercury gives a material resembling thick mercury. It is considered sodium oxide is formed which can be detected, indirectly, by heat of reaction measurement.
- A reduction in thick mercury formation rates by reducing the mini (sodium content of the amalgam leaving the denuder) to <0.0001% Na, achieved by 10 mm graphite packing in the denuder and sodium molybdate addition. It has been shown that caustic entrainment from the denuder is a function of the mini.
- Operation of cells with low mercury levels in pump tanks result in lower rates of thick mercury formation, and also, no thick mercury formed at the inlet end of the cells, indicating that excess water dispersion reacts with dehydrating agents (Sr amalgam and Na_2O). This prevents the formation of dry dispersions. Furthermore, an increase in fine water dispersions has been measured under these conditions.
- Operation of a converted J Unit Cell, with a mercury flow set at 120 l/min to 130 l/min, reduced the rate of thick mercury formation considerably, compared with the rates observed at 50 l/min. It is considered that at high mercury flows, the residence time on the baseplate is reduced for dehydration and subsequent gelling.
- Sodium chloride is not found in thick mercury. This suggests the aqueous dispersions present originate outside the brine cell.
- Measurements using tritium tracers have shown that the origin of the hydrogen atoms in the hydroxides, found in the thick mercury, is the caustic soda in the denuder or the water in the mercury pump tank.

The structure of the deposit is understood in terms of a colloidal gel with a dispersion of inorganic material, given in the above analysis, as fine particles amalgamated by mercury: their attraction for each other being stronger than their attraction for mercury so that they agglomerate, forming an open network with mercury in the capillary spaces.

As finely divided dispersions have a high surface energy there needs to be a process which transfers energy to the system. There are two probable sources of the sol precursor of thick mercury. These are:

- the reaction of dry amalgam with oxygen in air dispersions to give sodium oxide and mercuric oxide. There are several places in the cell where air and amalgam react, in the absence of water, under vigorous agitation conditions to give sodium oxide and mercuric oxide, for example the denuder lute where very weak amalgam leaves the denuder. Fine sodium oxide and mercuric oxide particles amalgamate and disperse in the amalgam together with traces of iron. Reaction of these dispersions with some water probably occurs, before the dispersion gels later on the baseplate.
- the formation of fine droplets of water and caustic soda, which later react with amalgam to form sodium and strontium salts. The fine droplets of water and caustic soda are broken to fine particles at the mercury pump. In the brine cell the aqueous dispersions dehydrate by reacting with sodium and strontium and become dry. These dry particles probably react with remaining air to give a gel which is rich in strontium and forms lower down the baseplate.

It is considered that the adhesion of the gel to the baseplate is due to the high magnetic susceptibility of the iron dispersions which are attached by weak magnetic fields at the baseplate. A model of the formation mechanism is given in Figure 2.

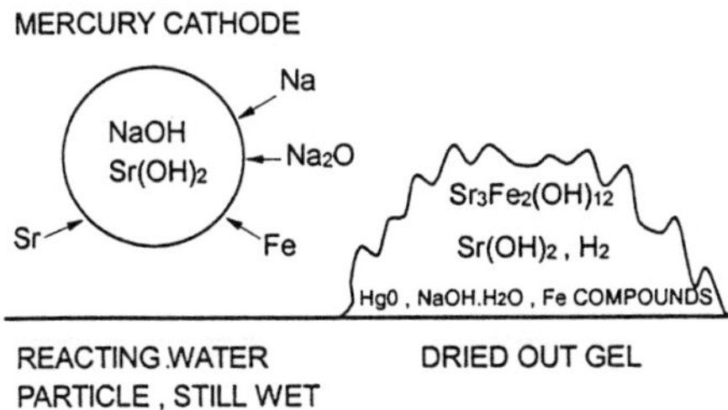

Figure 2 *Formation of Baseplate Thick Mercury*

4 METHODS TO SUPPRESS THICK MERCURY FORMATION

Various attempts were made to suppress the formation of thick mercury. These were based on the believed mechanism for the formation of the precursor sol and the drying process in the cell.

No practical methods were found to reduce the concentration of dispersions entering the cell to the required level. Effort to prevent the drying process in the cell by reducing the strontium level in the brine from 8 ppm to 2 ppm, by excess sodium carbonate treatment of the brine, did reduce the rate of thick mercury formation by at least one third. However this improvement did not justify the cost.

5 EFFECT OF PURE BRINE

The development of ion exchange technology for the membrane process has led to the production of very pure brines. The effect of a brine with very low impurity levels e.g. strontium level <50 ppb, using ion exchange columns filled with Duolite (ES467) resin, has been examined at two J Unit cells. At 7.6 kA/m^2, consistently low rates of thick mercury formation (mean <0.01 mm/day) were obtained at both cells.

The results are given in Figure 3. These show a substantial reduction in the peak rate, from approximately 0.7 mm/day to 0.1 mm/day. Although samples of thick mercury produced at the inlet end of the cell were not taken, it is presumed that the peak rate at anode cover 3 position was primarily due to Na_2O, HgO and NaOH precursor sol.

The trial showed that the formation rates only increased when there was a break through of strontium at the ion exchange column.

The trials also showed that operation of the cells at a reduced gap (~1.5 mm) was achievable without loss of current efficiency.

Currently for the waste brine process, there is no economic case for the reduction of thick mercury formation rates by the partial reduction of the strontium levels in feed brine, as obtained by the co-precipitation of strontium with calcium, using sodium carbonate. Also, an assessment of the benefits by the use of pure brine shows that there is no economic advantage for a feed brine ion exchange installation. This, of course, may change as ion exchange resins are developed further.

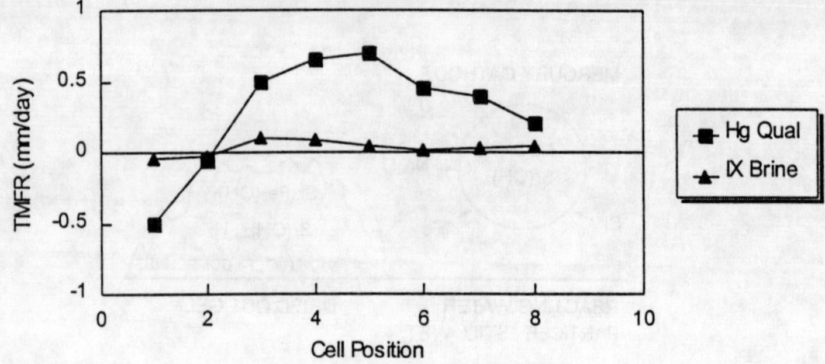

Figure 3 *Thick Mercury Formation Rates at 7.6 kA/m² (mm/day)*

6 CONTROLS

Unlike cell installations with automatic anode adjustment or where the rates of thick mercury formation are insignificant, the Castner Kellner manually adjusted cells are set to an overall k factor higher than the optimum for minimum power consumption and, as the mercury film increases, allowed to operate to a k factor which is close to the minimum power consumption. This operating k factor range depends on the manageable interval between cell treatments for elevated hydrogen levels in cell-gas, reduced mercury flows or low k factors. All these effects are due to a deteriorating cathode performance.

During the pure brine trials at the two J Unit cells, the unit was operated continuously at its load limit. Close monitoring of the unit parameters indicated a loss in unit current efficiency performance which was greater than that attributed to the increase in the cell gas hydrogen levels.

Mercury film thickness measurements at a number of cells showed that the peak rate of thick mercury formation was at the centre of the cells. The effect of this was not detected by the cell voltage measurements and anode shorts were occurring.

The merits of a k factor profile that anticipated the formation of thick mercury were clear. In order to check the consistency of the thick mercury growth patterns, k factor profiles at the end of life cells were obtained during high load operation. Approximately 67% of the cell checks showed the peak thick mercury formation rates were in the area of covers 4 and 5, although there was a variation in the actual rates. The peak rates for the remaining cells were at covers 6 and 7. As these cells were protected by the cell voltage measurements, it was decided that there was a case to modify the adjusting profiles. This was progressed at both J and K Units. The overall effect of the corrections was a unit current efficiency increase of approximately 1.5%, increasing the current efficiency from approximately 94.0% to 95.5%. The unit k factor increased from 0.153 to 0.157. The overall effect was a power consumption improvement of approximately 30 KW Hr/te Cl_2, at 8 kA/m².

A comparison of the effects on the end of life k factor profiles for anode covers 4 and 5, for the normal and modified k factor adjusting profiles, is shown in Table 1.

Table 1 shows that for the normal adjusting profile there were 19 shorting plates at a cell k factor of 0.133. The measured current efficiency was 91.7%. Also, for the modified profile the number of shorting plates was 3, at cover 5, at a cell k factor of 0.130 (Table 2). The measured current efficiency was 93.5%.

Table 1 *Effect of Normal Anode Adjusting Profile on Observed End of Life K Factor*

Cell K37E External Kf 0.133
Load (KA) 173 Feed Brine deg C 35.6

_____ Exit Brine deg C 79.6

k factor units = $Vm^2 kA^{-1}$

Measured C.E 91.7%

Observed K Factors At Anode Cover 4 Observed K Factors At Anode Cover 5
Shorting Plates (<0.070 K Factors) ▩

C	R	1	2	3	4	5	6	AVG
4	1	0.106	0.110	0.083				
4	2	0.081	0.079	0.070				
4	3	0.082	0.078					0.077
4	4	0.084	0.097	0.071				
4	5	0.088	0.090	0.091				

C	R	1	2	3	4	5	6	AVE
5	1	0.091	0.110					
5	2	0.091	0.099	0.100	0.092	0.077	0.080	
5	3	0.095	0.099	0.100		0.093	0.096	0.098
5	4	0.090	0.115	0.111	0.101	0.107		
5	5	0.125	0.107	0.121	0.137	0.108		

Table 2 *Effect of Modified Adjusting K Factor Profile on Observed End of Life Profile (-.015, +.000,+.025, +.020, +.015, +.010, +.005, .000)*

Cell K37E EXTERNAL Kf 0.130
Load (KA) 174 Feed Brine deg C 35.6

_____ Exit Brine deg C 79.3

Measured C.E 93.5%

Observed K Factors At Anode Cover 4 Observed K Factors At Anode Cover 5
Shorting Plates (<0.070 K Factors) ▩

C	R	1	2	3	4	5	6	AVG
4	1	0.128	0.131	0.127	0.119	0.088	0.071	
4	2	0.110	0.114	0.097	0.089	0.081	0.074	
4	3	0.121	0.104	0.089	0.089	0.087	0.084	0.101
4	4	0.129	0.132	0.100		0.096	0.093	
4	5	0.099	0.096	0.099	0.115	0.086	0.089	

C	R	1	2	3	4	5	6	AVE
5	1	0.090	0.099		0.078	0.071	0.083	
5	2	0.082	0.076	0.075	0.109	0.090	0.104	
5	3	0.076				0.077	0.081	0.084
5	4	0.075		0.079	0.075	0.080		
5	5	0.104	0.084	0.086	0.101	0.098		

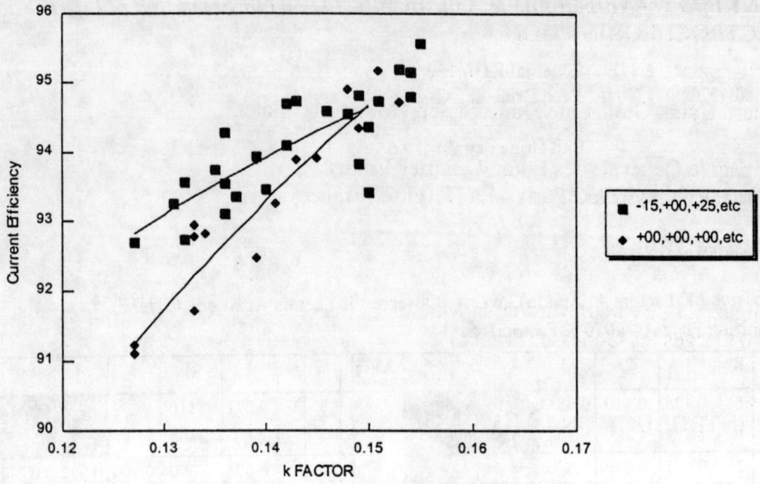

Figure 4 *Performance of current profile -15, +00, +25, +20, +15, +10, +05, +00*

Although the effect of the modified adjusting profile at the cells was clear, with respect to the reduction in the number of shorting anodes during their lives, proving the current efficiency benefits at the cells was not initially forthcoming due to the inadequacy of the measurement technique. This was improved significantly by the development of the amalgam concentration method for current efficiency measurement.

The current efficiency results obtained by the amalgam method during the life of the above profiled cell, k factors corrected to a standard temperature and load, are given in Figure 4.

The results show the current efficiency divergence as the cell k factor reduces, between the modified k factor adjusting profile and the unmodified adjusting k factor profile.

Experience has shown that it is prudent to monitor the thick mercury formation patterns, particularly when significant changes occur in the production rates.

ACKNOWLEDGEMENT

The laboratory work on the structure of thick mercury was carried out by D. Lee of Research Department. The operation of the J Unit cells with pure brine was carried out by K. Hancock of the Experimental Section.

23
DRINKING WATER PRODUCTION ON AN URBAN SITE: ELECTROCHLORINATION

Bernard Cyna, Guillaume Dourdin and Dominique Gatel

Compagnie Générale des Eaux, Quartier Valmy
32 Place Ronde, 92982 Paris - La Défense (France)

Bruno Langlais

Trailigaz, 29 - 31 boulevard de la Muette
95140 Garges les Gonesse (France)

1 INTRODUCTION

Drinking water treatment plants are often located in the centre of urban areas. In this type of environment, it is obviously impossible to stock explosive products. However, chlorine gas, used to disinfect water, can present a danger to the surrounding populations should there be an incident. An alternative now exists: electrochlorination.

2 THE DUTY TO DISINFECT

Disinfecting water means removing all the pathogenic organisms it contains to ensure its microbiological quality. This is an absolute necessity and the duty of all water producers. For nearly 100 years, thanks to disinfection using chlorine, water-borne illnesses have been progressively and largely eradicated. After its initial disinfection action, the chlorine retains its efficiency throughout the distribution system, the site where bacteriological developments occur.

Nonetheless, chlorine presents two major inconveniences. The first, well known by consumers, is the disagreeable taste of water when it is overchlorinated. The second is the appearance of organochlorinated compounds able to present long term health risks. Concerning these latter, a European directive is soon to be introduced that will limit their maximum admissible concentration.

These inconveniences have resulted in other treatments, such as ozonation and membrane filtration, being favoured. Although ozone is an excellent disinfectant and more efficient than chlorine, it does not leave any residual in the distribution system. Its action therefore needs to be completed by a low dosage chlorination to maintain the necessary residual level.

Membrane filtration processes represent a physical barrier that prevent the passage of micro-organisms. This particularly applies to reverse osmosis and nanofiltration. However, similarly to ozone, a slight chlorination is needed to complete the treatment and thus ensure a disinfectant residual in the distribution system.

Even though it is not the only disinfectant available, chlorine remains the most efficient way of maintaining the drinkability of the water through to the consumer's tap. It also remains the ultimate safety measure should the water resource become polluted.

The post-chlorination disinfection of water is generally carried out using either chlorine gas or a concentrated sodium hypochlorite solution, delivered and stored in the treatment plant.

Chlorine gas, because it is easier and less expensive to use, has frequently been the preferred solution.

However, Public Authority concerns regarding the safety of people have recently been translated by reinforcements to the regulations governing the storage of chlorine gas. Danger studies based on the simulated explosion of a chlorine gas tank have led to the definition of danger perimeters around installations which are incompatible with urban development.

Concerning the chlorination itself, the use of bleach concentrated at 150 g/l represents a standard alternative to chlorine gas. But bleach presents certain inconveniences. These include:

- major equipment corrosion problems leading to the installation having a reduced working life (5 to 10 years),
- the expense of the product,
- the concentration of sodium hypochlorite reduces rapidly over time and oxidation byproducts increase,
- risks to people resulting from interventions on the installation as well as risks linked to transporting bleach exist and should not be underestimated.

Electrochlorination is an attractive alternative.

3 THE ELECTROCHLORINATION PRINCIPLE

Electrochlorination consists of producing a bleach solution diluted on site by the electrolysis of a salt solution.

The principle of electrochlorination is the production of diluted sodium hypochlorite (NaClO with a concentration of 6 g/l) by the electrolysis of a sodium chloride solution.

Chlorides are oxidised into chlorine by anode: $Cl^- \rightarrow 1/2\ Cl_2 + e^-$
Sodium ions are reduced by cathode: $Na^+ + e^- \rightarrow Na$
with production of sodium hydroxide and hydrogen: $Na + H_2O \rightarrow NaOH + 1/2\ H_2$

When there is no diaphragm in the reactor, the chlorine formed near the anode reacts with the sodium hydroxide formed near the cathode to form sodium hypochlorite:

$$Cl_2 + NaOH \rightarrow NaOCl + HCl$$

Approximately 3.3 kg of salt and 5 kWh are required to produce 1 kg of chlorine in the form of 165 l of sodium hypochlorite.

4 ELECTROCHLORINATION: A LARGE NUMBER OF ADVANTAGES

This process has four main technical advantages:

- Corrosion problems associated with equipment are much reduced, the installation is clean and the working life is estimated as twice if not three times that of a concentrated bleach installation.
- Low concentration sodium hypochlorite is only stored for short periods, thus ensuring that its chemical characteristics and its quality are maintained. This means that fewer byproducts are formed over time when compared with commercial bleach.

- The operational costs of this type of installation are similar to those of a chlorine gas installation. This particularly applies to the variable costs: a kilogram of chlorine produced from concentrated bleach can cost 50% more than when produced on site using electrochlorination.
- Finally, electrochlorination removes the risks linked to the transport and handling of chemical products.

For several years, electrochlorination has been used in England to replace chlorine gas: the Chertsey plant near London produces 6.75 kg/h of chlorine whilst the Walton plant produces around 40 kg/h. In France, the town of Saint-Maur, whose water treatment plant is also located in an urban setting, was the first to decide to use this process. The installation has been operational since November 1993 and produces 6 kg/h of chlorine, being enough to satisfy the disinfection requirements of the plant which produces 50,000 m^3 of water per day.

The excellent results obtained have led to this same principle being adopted for the Choisy-le-Roi water treatment plant, which has a nominal production rate of 800,000 m^3/day. The electrochlorination unit produces 37 kg/h since the end of 1995 (Figure 1).

Figure 1 *The Choisy-le-Roi electrolysis unit*

5 OPERATION OF THE CHOISY-LE-ROI ELECTROCHLORINATION INSTALLATION

The foodstuff quality salt is mixed with softened water in a salt tank until a saturated solution is formed.

The saturated brine is then diluted in softened water to obtain a solution with a 3% concentration. This solution is then fed to the electrolyzer unit, constructed from several series of anode/cathode cells, and exposed to a direct current charge. The result is the production of sodium hypochlorite and hydrogen.

This mixture is then fed to a storage tank where the separated hydrogen gas is immediately diluted with the air using a forced ventilation system and then discharged to the atmosphere at a very low concentration level (<1% of the explosive dose). A danger study has shown a complete absence of any safety risk.

The resulting sodium hypochlorite solution, with an average concentration of 6 g/l, is then injected into the water at the different existing chlorination points. The softened water is produced on site using an ion exchange resin system. The installation is entirely automated and supervised from the command station (Figure 2).

The electrolysis operates on an 'all or nothing' basis at its maximum flow rate. One of the storage tanks is reserved for commercial bleach used to make up for any insufficiencies and as a standby source.

The composition of the sodium hypochlorite solution has been analysed. At a maximum chlorination dose, the quantity of impurities introduced into the water remains lower than one tenth of the maximum acceptable concentration.

Figure 2 *The electrochlorination installation. Foreground: the three hypochlorite storage tanks; middleground: the hypochlorite production building; background: the two dissolver silos*

6 CONCLUSION

Electrochlorination represents a clean and ecological technology which presents all security guarantees both inside the plant and for the surrounding population. Drinking water producers in England, and now in France, are increasingly adopting this process. Electrochlorination represents a solution that looks to the future of drinking water treatment plants located in an urban environment. Its use is also perfectly adapted to small chlorination stations located along the distribution system.

24

PROCESS ENGINEERING CONSIDERATIONS IN CHLORINE COMPRESSION AND LIQUEFACTION

T. F. O'Brien and I. F. White

Raytheon Engineers and Constructors
30 South 17th Street,
Philadelphia, PA 19103, USA.

Raytheon Engineers and Constructors
C I Tower, St.George's Square
New Malden, Surrey, KT3 4HH, UK.

1 INTRODUCTION

The final step in the production of merchant chlorine is its liquefaction. Even when chlorine is to be consumed on site, liquefaction is often used to provide a buffer between processes or to reject harmful impurities. When chlorine is produced at low pressure, compression always precedes liquefaction in order to allow condensation at reasonable temperatures. This chapter considers the unit operations of compression and liquefaction.

2 COMPRESSION

2.1 General Considerations

Thermodynamically, the compression of chlorine resembles that of air and most other diatomic gases. Specific heat ratios vary little, and so the energy required for a given compression ratio and the resulting rise in temperature are quite similar. Reduced pressures of the compressed gas usually are less than 0.1, so compressibility factors are close to unity and pressure-volume relationships nearly ideal. Some points which might be considered "special" are:

1) A release of gas can have serious consequences due to toxicity and corrosivity.
2) Moisture must be effectively excluded to prevent corrosion of the compressor and downstream equipment.
3) At sufficiently high temperatures, chlorine reacts combustively with many metals, including the ferrous types commonly used in compression systems.
4) Great care must be taken to prevent contact between the gas and common types of lubricant, in order to avoid the loss of lubricating power and even the risk of fire.
5) Backflow must be prevented at shutdown, in order to prevent releases from upstream equipment and reaction of dry chlorine with titanium in the cooling system.
6) Ingress of air or seal gas will reduce the efficiency of the liquefaction process which usually follows compression.

2.2 Centrifugal Compressors

Centrifugal compressors are the modern industry standard in larger plants. They are designed for large flows, and larger sizes are more efficient. Because they resemble centrifugal pumps in first producing a high velocity in the fluid and then converting kinetic energy into static pressure, centrifugal machines also operate more efficiently on heavier gases such as chlorine. Start-ups present a problem when the low-pressure piping and equipment are filled with air; compressor characteristics will reflect the low molecular weight and will change as chlorine displaces the air. This situation must be considered in the design of a compressor and its controls.

2.2.1 Specific Safety Measures. We have already summarised the major safety considerations in chlorine compression.[1,2] Some of the specifics are repeated below.

Because chlorine can react combustively with steel, it is essential to limit temperatures within the system. This is done in the first instance by limiting the compression ratio achieved in each stage and providing cooling of gas between stages. Machinery should be designed to avoid "hot spots", or inserts of metal more resistant than iron or steel should be used where temperature is expected to be high. Final protection is through a shutdown initiated by any of a number of high-temperature switches.

Backflow of compressed gas after shutdown is especially likely with centrifugal compressors. Some form of mechanical prevention is necessary. We prefer the use of automatic valves which close when differential pressure is reversed to the less reliable non-return valves.

Before compression or between stages, the gas is often cooled by direct contact with liquid chlorine recycled from storage or the liquefaction plant. This approach makes use of the latent heat of vaporisation and allows cooling to quite low temperatures. It is important to avoid excessive entrainment of liquid into the compressor itself. There is also the risk of accumulation of liquid in the cooling apparatus. This is hazardous because of the selective condensation of nitrogen trichloride. NCl_3 is a highly unstable and dangerous compound which results from the chlorination of nitrogen compounds. Its decomposition has resulted in a number of serious accidents and even fatalities. Control of this hazard is beyond the scope of this chapter, but much of the literature on the subject has been assembled by The Chlorine Institute.[3]

2.2.2 Seals. Preventing releases of chlorine is of paramount importance, and the sealing system must be of high quality. Gas-loaded labyrinth systems are quite common. For safe operation, the seal gas must be reliably very dry and at sufficient pressure to ensure that any leakage will be into the process. For efficiency in liquefaction, the rate of flow of gas should be quite small. High temperatures resulting from friction at sealing points have been a frequent cause of fire.[4] Some of today's designs use stationary abradable polymer seals which prevent metal-to-metal contact.[5] They also keep the leakage path narrow and so reduce the rate of flow of gas through the seal.

2.2.3 Surge Control. The phenomenon of surge is a consideration in the design of centrifugal systems. If gas throughput is too low, flow patterns become unstable and the delivered head begins to decrease. This allows a reversal in the direction of flow, followed by rapid oscillation and perhaps severe vibration (surge). All centrifugal installations therefore include a provision for recycle of compressed gas to the suction of the compressor. Regardless of net forward flow, the total flow through the compressor then can be kept above the surge point.

Recycle of gas increases the energy expended in compression. To minimise this effect while maintaining safe operation, surge control systems can become quite sophisticated. The systems usually met in chlorine applications are less so. One reason for this is that most chlorine compressors are not impressively large and the potential for reduction in energy consumption is small. There are also process-related reasons.[6] For example, the compressor suction pressure control commonly is also responsible for maintaining the cell output pressure. It is more important to keep precise control of this (low) pressure than to save a small fraction of the compression energy. It is also good practice to maintain some recycle even when not strictly needed for surge control. This allows faster response when needed for safety reasons.

2.3 Reciprocating Compressors

Multistage reciprocating compressors are not the modern choice for primary compression, partly because of their high maintenance requirement and sensitivity to entrained liquid or solid. Compression between stages of liquefaction, however, is a very different matter. The suction pressure is higher, the gas flow is much lower, and there is less chance of a problem with dirty gas. Reciprocating machines are therefore often chosen for this application.

Unless quite small, reciprocating compressors should be equipped with pulsation dampeners or surge tanks with volumes greater than four times the maximum displacement volume of the compressor chambers. To prevent contact of chlorine with lubricating oil, we recommend the use of double distance pieces. These should be purged with dry gas. On the process end, this gas should be at a higher pressure than the chlorine. There will be some leakage into the process, and this must be recognised in downstream use or processing.

2.4 Rotary Compressors

Liquid-ring compressors are our main concern in this section. They are limited in size, and so we consider them to be suitable only for smaller plants. Other disadvantages are low energy efficiency and high maintenance cost. Provision of installed spare units should be considered when several machines operate in parallel or when reliability of output must be very high.

A compensating advantage is that the liquid absorbs much of the heat of compression. The exit gas temperature is then less likely to limit the compression ratio. Compression ratios can be five or more, and so only one stage of compression is required before liquefaction.

Since the liquid and gas leave the body of the compressor together, a collector/separator is installed close to the outlet. When the liquid is sulphuric acid, some sort of mist eliminator should be added. Waste heat is removed by cooling the recirculating acid. This can be pumped, or the pressure developed by the compressor can be used to return it to the suction. With use, the acid will become more dilute and more corrosive. Its strength should be kept above 94% by replacing spent acid with fresh material. Disposal of the spent acid may be a problem; it will contain dissolved chlorine and possibly impurities from the low-pressure chlorine process.

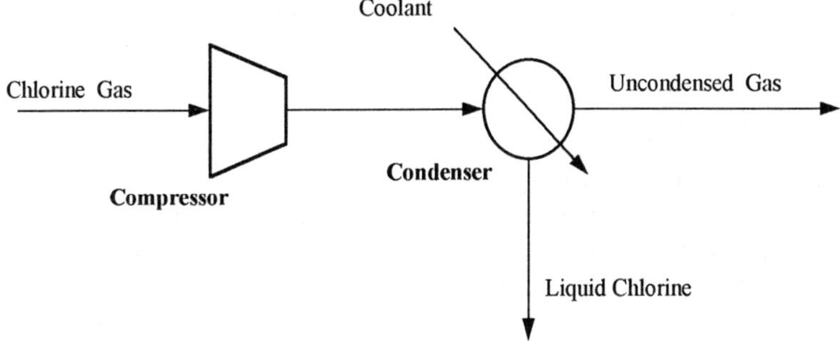

Figure 1. *Single Stage Liquefaction of Chlorine*

3 LIQUEFACTION

3.1 General Considerations

The simplest approach to compression and liquefaction is the one-stage process shown in Figure 1. The gas is brought from approximately atmospheric to an elevated pressure, and chlorine is condensed by a cooling medium.

Chlorine as manufactured is always accompanied by non-condensible gases (referred to below as "inerts"). There are several sources of these gases:

1) oxygen and hydrogen produced by current inefficiencies in the cells
2) air and carbon dioxide contributed by the brine
3) air leakage into subatmospheric parts of the process
4) air or nitrogen applied to compressor seals
5) air or nitrogen used to transfer liquid chlorine
6) gas displaced from returned shipping containers
7) air unintentionally introduced during evacuation of equipment
8) air or nitrogen used to purge systems before opening for maintenance
9) air displaced from piping and equipment during start-up
10) air or nitrogen added to control hydrogen content in liquefaction
11) hydrogen produced at cathodic spots or leaked from cathode chambers

The inerts pass through liquefaction and are vented from the system. Chlorine is present in the vent at a partial pressure equal to its vapour pressure over the condensed liquid. Complete liquefaction is therefore not possible, and the degree of liquefaction depends on the temperature and pressure at the end of the process as well as on the purity of the chlorine feed.

3.2 Achievable Degree of Liquefaction

Using a simple mass balance and the close approximation that the chlorine content of the gas leaving a liquefier corresponds to the vapour pressure of the liquid at the operating temperature, the fraction of the incoming chlorine removed from a gas is given by equation (1).

$$L = 1 - \frac{p\,(1 - y)}{y\,(1 - p)} \qquad (1)$$

where p = ratio of vapour pressure of chlorine to total pressure
y = mole fraction of chlorine in feed gas

Efficient liquefaction results when p is low and y close to unity.

3.2.1. Gas Purity. The parameter y is the purity of the gas fed to liquefaction and will be high when the cell gas is pure and not diluted during processing. Pure cell gas results when current efficiency is high and the brine does not contribute diluents to the gas. High current efficiency depends on proper operating conditions, proper care of the anode processes, and efficient operation of the anolyte/catholyte separator (e.g., the membrane). Brine will contribute diluents if it is highly aerated or contains excess carbonate ions added during its purification. These ions will react with the chlorine produced at the anode to form carbon dioxide, which dilutes the gas. This can be avoided by acidifying the brine to destroy the carbonate and removing the resulting CO_2 before entry to the cells. Excess acid will then react with hydroxide ions which migrate from the catholyte, reducing the amount of oxygen formed.

In-process dilution can result from infiltration of atmospheric air or compressor seal gas or from the deliberate addition of an inert to control hydrogen concentration. To prevent corrosion, it is especially important to avoid the ingress of damp atmospheric air. This is accomplished by maintaining a tight system and keeping the gas pressure positive or only slightly negative at the compressor suction. With labyrinth seals on a compressor, there will be a certain amount of flow into the process gas, but with good design this flow can be kept small and dry.

Transfer of chlorine can add inerts to the liquefaction system. Liquid is often sent from storage to users or shipping containers by added gas pressure. When a storage tank is nearly empty, it must be prepared to receive new product by venting this gas through a scrubber or the liquefaction plant. Similarly, accumulated impurities must be disposed of when shipping containers are returned, and frequently a container must be evacuated for maintenance or revalving.

Other maintenance activities, which include evacuation of systems and purging of piping and equipment, also contribute inerts. Purging is also necessary at start-up, and liquefaction efficiency will be low until steady-state conditions are approached. Cellroom start-ups must be managed under this constraint.

As chlorine condenses, the mole fractions of more volatile components increase in the gas, and the concentration of hydrogen may exceed its lower explosive limit of about 4% by volume.[7] While some systems are designed to contain a low-level hydrogen explosion, a more common practice is to add nitrogen or dry air whenever this can occur. This is often the major source of dilution, and its use reduces the incentive for elimination of other sources of inerts.

3.2.2 Temperature and Pressure. The parameter p is small when the total operating pressure is high and the vapour pressure of chlorine low. Both these conditions are readily achieved in principle, but the energy consumption and the cost of plant both increase rapidly as a result.

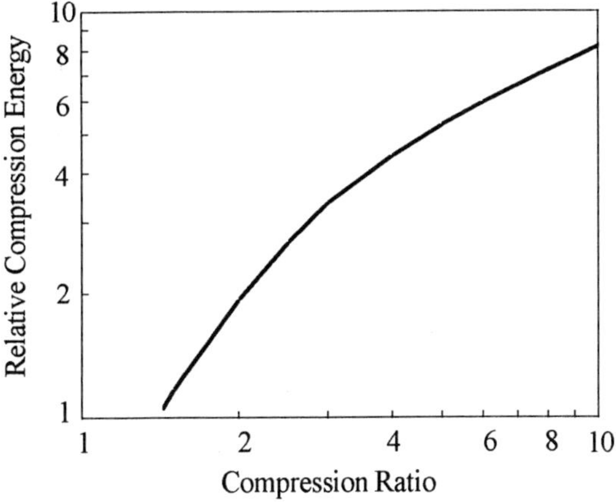

Figure 2. *Energy Consumption in Compression*

A high compression ratio contributes to a low value of p. Figure 2 shows the energy required in compression as a function of the compression ratio. Figure 3 deals with the other side of the process. The abscissa is the reciprocal of the vapour pressure of chlorine. Again, a high value of the co-ordinate is desirable. It is noteworthy that over their usual ranges these logarithmic curves are nearly parallel. Minimisation of the total energy consumption therefore implies approximately equal energy expenditures in compression and liquefaction. In our experience, most practical compression/liquefaction systems follow this crude rule of thumb.

Figures 2 and 3 also suggest the common approach of staging the process. Most of the chlorine can be recovered, and most of the energy consumed, under conditions less rigorous than required for the desired degree of liquefaction. The cost of this energy transfer is correspondingly low. Liquefaction then is completed under the more severe conditions, but the amount of gas processed and the amount of energy transferred under those conditions are reduced.

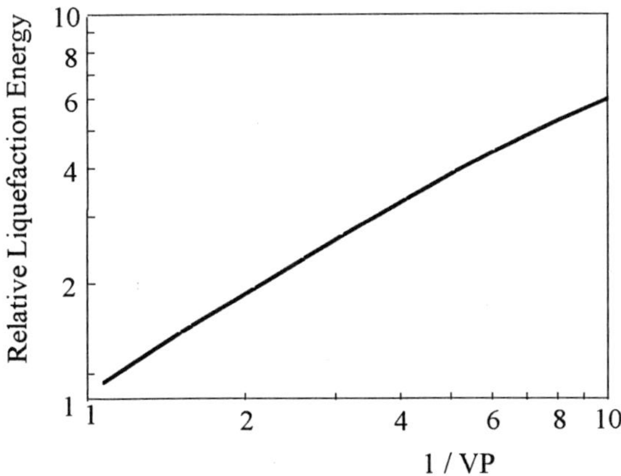

Figure 3. *Energy Consumption in Liquefaction*

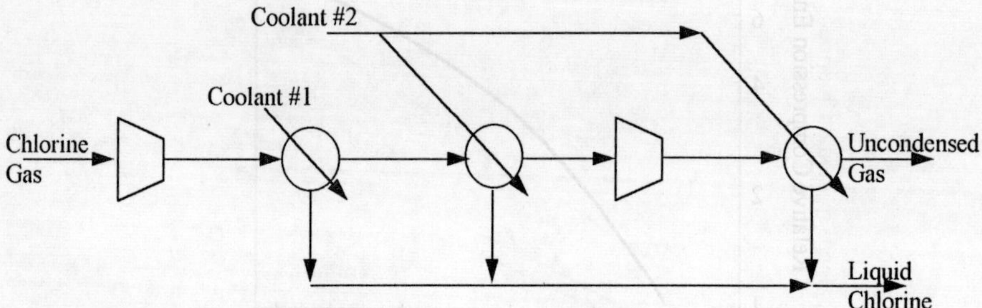

Figure 4. *Three-Stage Liquefaction*

In practice, liquefaction systems tend to be limited to three discrete stages which progress from relatively high temperature and low pressure to low temperature and high pressure. Both chlorine and refrigerant are then applied at two different pressures. A typical scheme is shown in Figure 4. Two-stage systems in which only one variable is staged are quite common, and a typical arrangement would be that without the last stage shown in the figure.

The trade-off between temperature and pressure raises the question of the use of cooling water rather than a compressed refrigerant. Since at 35 °C, chlorine has a vapour pressure of about 1000 kPa, this is not a widespread practice. Pressures of 1200-1600 kPa would be required to achieve 80% liquefaction. Most plants find it practical to operate primary liquefiers in the range 300-450 kPa and to condense chlorine with a refrigerant.[8] If cooling water is reliably at a comparatively low temperature or if a chlorine-containing gas is already at an elevated pressure, the economic/technical balance can change in favour of cooling water.

3.3 Inerts in Liquid

3.3.1 Accumulation. When a very high degree of liquefaction is necessary, we are faced with the possibility of accumulating excessive concentrations of "noncondensibles" in liquid chlorine. With 99.5% liquefaction, for example, an inert comprising 3% of the feed gas and with a relative volatility of 1000 will have a liquid-phase concentration of about 0.5% (m/m). In the more reasonable two-stage approach in which each stage liquefies about 93% of the incoming chlorine, such a concentration exists only in the liquid from the second stage and the overall concentration is less than 0.1%. This will more than double if liquefaction is pushed to 99.9%.

The rapid increase in the concentration of dissolved inerts at high degrees of liquefaction is perfectly analogous to the increase in energy consumption. Staging the process again makes a dramatic difference. Figure 5 shows typical data for 90 to 99.9% liquefaction in one, two and three stages.

While many electrolytic cell gases can therefore be nearly completely liquefied without high concentrations of dissolved inerts, other sources of chlorine which may become important in future can present problems. Deacon-process gases resulting from the oxidation of HCl, for example, will have much higher inerts concentrations. Good results then are obtained in conventional liquefaction only by reducing the value of p accordingly.

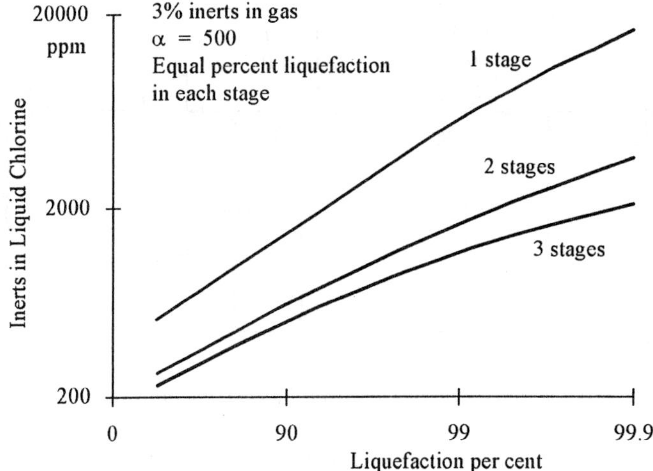

Figure 5. *Effects of Process Staging on Dissolved Inert Concentration*

This is expensive and will give a product with more than normally specified levels of dissolved inerts. In the 99.5% two-stage case quoted above, a gas with 50% inerts would produce a liquid with an inert concentration of 2.5%. Recent patents and an economic evaluation[9] therefore consider partial liquefaction followed by absorption of the residual chlorine. Carbon tetrachloride is the usual example of an absorbent. Physically, it is ideal for the application and so historically has been used in sniff gas recovery systems. Because of environmental restrictions, some of these systems have been removed from service. CCl_4 is disappearing as an item of commerce and would not seem a practical choice for a new process.

3.3.2 Removal. When bulk storage is provided for chlorine, the tanks will be equalised with the liquefiers and some of the dissolved inerts will incidentally weather off. This is not suggested as a method for their removal, but it is a consideration in design and rating of the liquefiers and vent gas system. A more positive process is the removal of the inerts by partial vaporisation in a column.[9] We have devised a simpler approach giving much the same effect by flashing liquid back to a lower pressure. This is conveniently done in a multistage system by returning the vapour from a flash tank to the suction of the preceding compressor. Figure 6 illustrates this for a three-stage system in which the second is operated at a lower temperature than the first and the third at a higher pressure. Vapour from the last flash of the product should be returned at some intermediate pressure at which adequate flashing results while the storage pressure remains conveniently high. This flash of the primary liquid is the most effective step in reducing the total inerts concentration. When the vapour is returned to an intermediate stage of the primary chlorine compressor, the mechanical design of the machine and the characteristics of the surge-control system must reflect the change.

3.4 Water in Liquefaction Systems

3.4.1 Phase Equilibria. In any real system, the chlorine gas entering liquefaction will contain a small amount of water. Accumulation of water can result in corrosion and the deposit of solid hydrate of chlorine. This is a clathrate containing about eight molecules of

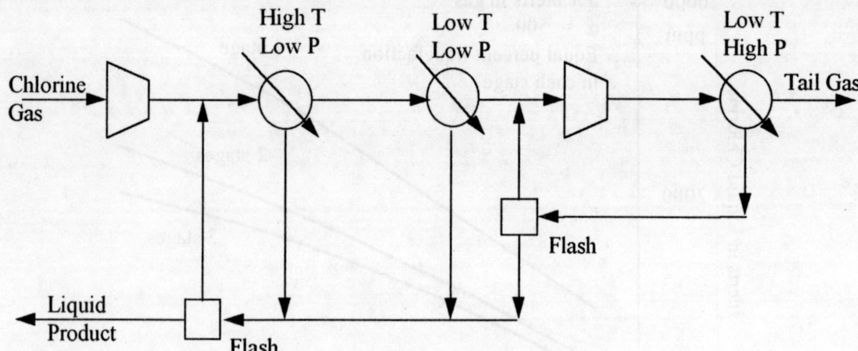

Figure 6. *Process for Reduction of Dissolved Inerts Concentration*

water to one of chlorine. At atmospheric pressure, it has a decomposition temperature of 9.6 °C. The quadruple-point temperature above which the hydrate does not exist is 28.7 °C.

During liquefaction, water partitions in accord with its equilibrium with chlorine. Because of its incompatibility with liquid chlorine, water has a very high activity coefficient. It behaves as a volatile compound relative to chorine and tends at first to gather in the vapour phase.

Experimental data on the chorine-rich side of the chlorine/water system are scarce. By combining a series of ingenious approximations with the behaviour of a model system, Ketelaar[10] was able to estimate the solubility of water and the vapour pressure of it or the hydrate over a range of temperatures. The relative volatility of dissolved water to chlorine was predicted to be nearly constant (about 4.2 at liquefaction temperatures). Other results can be summarised by the following equations, with solubility as the mole fraction and vapour pressure in atmospheres:

$$X_s = 426.6\ e^{-3880/T} \tag{2}$$

$$P_o = 4.51 \times 10^7\ e^{-6300/T} \tag{3}$$

3.4.2 Examples. The first liquid phase formed during liquefaction will be a homogeneous solution, and water will distribute according to its relative volatility of 4.2. The condensation of water will lag behind that of chlorine (Figure 7). When water reaches its solubility limit at the operating temperature, a second condensed phase forms and Figure 7 no longer applies. Water then will condense until its partial pressure in the gas equals the vapour pressure of the new phase. Under typical liquefaction conditions, the new phase will be the solid hydrate.

All the calculations which form the basis for this section use the estimates first made by Ketelaar. While those result from thermodynamic calculations without experimental verification, they are not without backup by real-world experience. Updyke[11] used very similar methods to predict the distribution of water in operating systems.

One system was a two-stage process in which 10 °C cooling water was used on the first stage and flashing chlorine on the second. From the first stage, calculations predicted a homogeneous solution in which all condensed water was soluble. There was in fact no corrosion. On the process side of the secondary condenser, the water content exceeded its

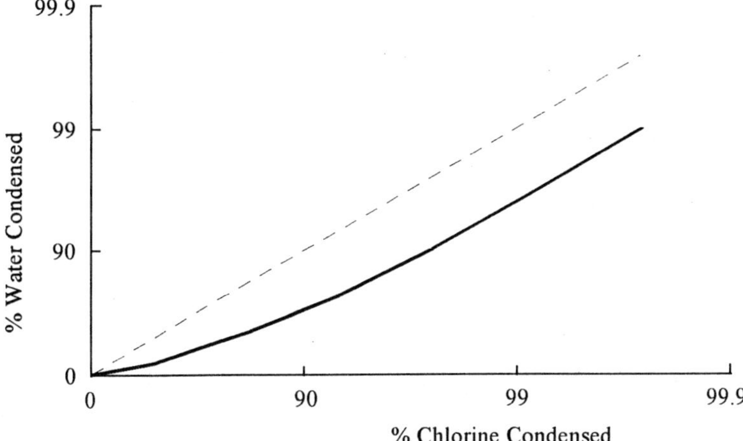

Figure 7. *Mutual Condensation of Water and Chlorine*

solubility and a second phase formed. So long as the hydrate was able to leave with the chlorine and did not accumulate, again there was no corrosion problem. The flashing side of the secondary condenser was another matter. Equilibrium calculations suggested that the saturated vapour would not hold all the water introduced with the liquid and that as a result hydrate would continuously accumulate in the exchanger. Indeed, the system proved troublesome and, because of melting of the hydrate, corrosion was severe during shutdowns. This study thus qualitatively confirms equilibrium calculations based on Ketelaar's method and demonstrates its utility.

We have calculated the results of liquefaction of a number of mixtures under different conditions. A key assumption, as shown in Table 1 for 99% liquefaction of a typical chloralkali gas, always is the water content of the feed gas. Reducing this concentration will often eliminate a hydrate problem. This emphasises the importance of proper design and operation of the cooling and drying processes which precede chlorine compression.[12] It also suggests that the common design technique of providing some redundancy in the drying system is a wise one.

4 TRENDS

4.1 Deep Liquefaction

In the typical chloralkali plant, liquefaction is followed by another process which removes chlorine from the tail gas. Examples are the absorption of chlorine in a solvent, the production of HCl by combustion with hydrogen, and the production of hypochlorite

Table 1 *Relationship Between Water Content of Gas and Formation of Hydrate*

ppm H_2O in dry gas	20	30	40	50	100
hydrate formation, kg/1000 te Cl_2	0	8	17	26	70

bleaches by reaction with alkaline compounds. These processes have become unacceptable environmentally or less reliable economically. Liquefaction systems in several plants have recently been supplemented to increase their recovery of chlorine. This trend will continue and will increase the importance of some of the points made above.

4.2 Other Sources of Chlorine

Changes in market conditions and the structure of the chloralkali industry have recently increased interest in the recovery of the chlorine value of by-product HCl.[13] Some techniques give a crude product very different from chloralkali cell gas. These suggest alternative processes which are beyond the scope of this chapter.

Chemical oxidation of HCl produces high concentrations of inerts. When the gas is under pressure, much of the unreacted acid and by-product water can be removed by ordinary cooling water. If the process temperature remains above the quadruple point of the chlorine:water system, hydrate cannot form. Liquefaction of the dried gas then follows the rules already discussed. The practical differences result from the high concentration of inerts. First, deep liquefaction will result in concentrations of dissolved inerts much higher than those associated with normal cell gas. Secondly, the reduced mole fraction of chlorine in the gas means that more water will condense at a given degree of liquefaction, and more hydrate will form.

A recent variation uses a two-bed system with recirculation of catalyst between the beds.[14] Control of the temperatures and of the oxygen and chlorine contents of the catalyst in each bed allows higher conversions of HCl and greatly simplifies downstream processing. The use of less excess oxygen also gives a more concentrated dry gas and relieves the problems noted in the preceding paragraph.

A more conventional process for recovery of chlorine values is the electrolysis of aqueous solutions. This is practised in a number of plants and can benefit from developments in membrane technology.[15] The gas is not radically different from that produced in a chloralkali cell, and the added complication of high levels of inerts associated with oxidation processes does not occur. An intriguing variation is the production of dry gas by electrolysis of anhydrous HCl.[16] While full information has not yet been published, one again would expect the gas to be quite similar to chloralkali process gas with the added benefit of a low water content.

REFERENCES

1. I.F. White, G.J. Dibble, J.E. Harker and T.F. O'Brien, 'Modern Chlor-Alkali Technology', ed. K. Wall, Ellis Horwood, Chichester, 1986, Vol. 3, Chapter 8, p.97.
2. I.F. White, 'Chloralkali Plants - Designing for Safety', Proceedings, Seminar on Essential Tools for Designing Safe and Reliable Plant, IMechE, London, 1993.
3. 'Nitrogen Trichloride - A Collection of Reports and Papers', Member Information Report 21, Edition 4, The Chlorine Institute, 1996.
4. E.S. Sokol, 'Chlorine Compressor Hazards - Fire, Oil, Water and Air'. Proceedings, 15th Plant Operations Seminar, New Orleans, The Chlorine Institute, 1972.
5. W.G. Hoppock, 'Centrifugal Compressor Revamps', Proceedings, Rotating Machinery Users Council Meeting, Long Beach, 1990.

6. T.A. Weedon, Jr., 'Pressure Control in Chlorine Plants', Electrode Corporation Chlorine/Chlorate Seminar, Cleveland, 1989.
7. 'Explosive Properties of Gaseous Mixtures Containing Hydrogen and Chlorine', Member Information Report 121, Edition 1, The Chlorine Institute, 1977.
8. T.A. Liederbach, Electrode Corporation Chlorine/Chlorate Seminar, Cleveland, 1991.
9. V. Wong and S-H. Wang, 'Chlorine from Hydrogen Chloride by the Carrier Catalyst Process', SRI International PEP Review 94-1-3, Menlo Park, 1994.
10. J.A.A. Ketelaar, 'The Drying and Liquefaction of Chlorine and the Phase Diagram Cl_2-H_2O', reprint distributed by The Chlorine Institute.
11. L.J. Updyke, 'Method for Calculating Water Distribution in a Chlorine Condensing System', Proceedings, 25th Plant Operations Seminar, Atlanta, The Chlorine Institute, 1982.
12. T.F. O'Brien and I.F. White, 'Modern Chlor-Alkali Technology', ed. R.W. Curry, Royal Society of Chemistry, Cambridge, 1995, Vol. 6, Chapter 7, p. 70.
13. T.F. O'Brien, 'Recovery of Chlorine from By-product HCl', Proceedings, Chloralkali Industry Update Conference, Philadelphia, Consulting Resources Corporation, 1996.
14. H.Y. Pan, T.T. Tsotsis, S. Benson and R. Minet, 'Process for Converting Hydrogen Chloride to Chlorine', *Ind. Eng. Chem. Res.*, 1994, **33**, 2996.
15. K. Schneiders and C. Herwig, 'Recycle of HCl to Chlorine', Proceedings, 38th Plant Operations Seminar, Houston, The Chlorine Institute, 1995.
16. D.J. Eames and J. Newman, 'Electrochemical Conversion of Anhydrous HCl to Chlorine Using a Solid-Polymer-Electrolyte Electrolysis Cell', *J. Electrochem. Soc.*, 1995, **142**, 3619.

25

A NEW METHOD TO REMOVE SODIUM SULPHATE FROM BRINE

Kenneth Maycock, Zbignew Twardowski and Judith Ulan

Kvaerner Chemetics Inc.
Vancouver, B.C., Canada

1 INTRODUCTION

The sulphate ion can have a detrimental effect on ion exchange membranes used to produce chlorine and sodium hydroxide. Manufacturers of membranes impose tight limits on the amounts of sulphate ion tolerated in the brine fed to the electrolyzers. As a large amount of the brine is recycled from the electrolyzer back to the saturation area, the sulphate ions which are dissolved, build up quickly in concentration and exceed the allowable limits. A number of processes have been designed to separate the sulphate ion from brine, most of these methods are either high in capital cost or high in chemical cost.

Kvaerner Chemetics Inc. has developed a method that is neither high in capital cost nor high in operating costs. The process uses a new technique known as nanofiltration, patents have been applied and granted for this new process.[1]

2 KVAERNER CHEMETICS PROCESS

The presently available methods for the control of sulphate level in the chloralkali brine circuit include:

- chemical precipitation as barium sulphate or calcium sulphate
- ion-exchange (e.g. NDS process)[2]
- crystallization of Na_2SO_4-10 H_2O (chlorate process only)[3]
- purge (sometimes with evaporation to reduce salt loss)

It would appear that purging of brine is the most common method of controlling the sulphate level in brine. All remaining process options are regarded as either too costly and/or cause operational problems in the plant.

Kvaerner Chemetics has developed a new method of separation of sodium sulphate from brine which incorporates the membrane nanofiltration technique, somewhat similar to the more widely known RO (reverse osmosis) membrane technique.[4,5] Figure 1 shows, schematically, the known high pressure membrane filtration techniques and their respective separation characteristics. It has been known that NF (nanofiltration) membranes can effectively separate multivalent anions (e.g. SO_4^{2-}) from monovalent ones (e.g. Cl^-) in dilute solutions. However, the anticipated loss of this selectivity and the high osmotic pressure necessary have prevented adaptation of the NF method for treatment of concentrated electrolytes. Kvaerner Chemetics has discovered that the high sulphate ion

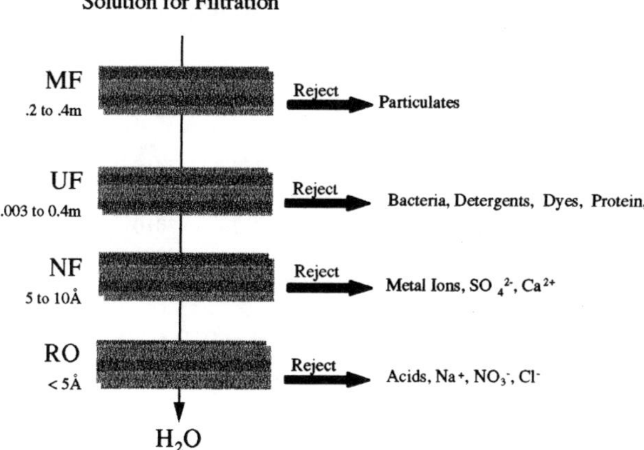

Figure 1 *Rejection characteristics of membrane filtration techniques*

rejection by NF membranes not only holds in concentrated brines but the selectivity is further enhanced by zero or even negative rejection of chloride ion species. The latter observation has an additional important benefit of minimizing the osmotic pressure difference between the feed treated brine and the "sulphate-free" brine permeate, that one would otherwise expect. As a result we have found that effective separation of sodium sulphate from chloralkali brine can be accomplished at feed pressures in the range of 2000-4000 kPa, which are very typical in operation of NF systems.

It is generally known that high pressure membrane processes such as MF (microfiltration), UF (ultrafiltration), NF or RO, require elaborate feed pretreatment to minimize membrane fouling by e.g. suspended solids. Here again, we have found that no pretreatment is required if the stream to be treated is either spent brine or re-saturated brine after the standard brine ion exchange purification.

We have also done preliminary experiments to verify whether other impurities in brine could be removed by the NF membrane method and have obtained some promising results. Tests were also done to demonstrate that the new technique can also be used to separate sulphate from the chlorate liquor, however a parallel separation of sodium chromate also occurs in this application.

3 LABORATORY PILOT PLANT TRIALS

The laboratory pilot plant work was carried out using a single membrane module of a standard industrial size. Results of the test runs are shown in Table 1 and Figure 2.

$$\%\ \text{Rec (Recovery)} = \frac{\text{Volume permeate}}{\text{Volume feed}} \times 100\% \qquad (1)$$

$$\%\ \text{Rej (Rejection)} = \frac{(\text{Conc in feed - conc in permeate})}{\text{Conc in feed}} \times 100\% \qquad (2)$$

Table 1 shows the sulphate rejection of our nanofiltration membrane at different sodium chloride concentrations and Figure 2 is a graph of sulphate rejection vs. sodium chloride

Table 1 *Effect of NaCl concentration on sulphate rejection**

Expt. #	Feed (gpl)		Concentrate (gpl)		Permeate (gpl)		Pressure	% Rec.	% Rej.	
	Na_2SO_4	NaCl	Na_2SO_4	NaCl	Na_2SO_4	NaCl	(kPa)		Na_2SO_4	NaCl
1	9.2	155.8	10.8	157.2	0.2	149.0	1410	25.8	98.2	4.4
2	11.3	189.6	11.9	184.2	0.1	177.9	2020	12.5	99.6	6.0
3	23.0	249.2	36.8	249.2	0.6	252.7	3010	11.8	98.9	-0.5
4	10.3	309.2	10.9	307.2	0.1	296.6	3800	9.5	99.3	4.1

*Single pass experiments, Temperature = 40 °C, pH = 9, 2.5 " X 40 " membrane module, total membrane area 1.7 m^2.

concentration. Sulphate rejection is greater than 98% over the entire range of sodium chloride concentrations studied. The sodium chloride rejection is very small, consequently the permeate that is returned to the process has lost very little salt.

Because sulphate rejection is independent of sodium chloride strength, the sulphate removal system can be applied to processes with widely different salt concentrations. In Table 1; Condition 2 corresponds to sulphate removal from weak brine in the membrane chloralkali process. Condition 3 corresponds to sulphate removal from the mercury chloralkali process. Condition 4 corresponds to sulphate removal from saturated brine.

Table 2 shows the sulphate rejection of a nanofiltration membrane at different feed sodium sulphate concentrations and Figure 3 is a graph of sulphate rejection vs. feed sulphate concentration. Increasing the feed sulphate concentration to as high as 70 gpl does not significantly decrease sulphate rejection. Therefore the sulphate removal system can be designed as a multi-stage system ultimately concentrating the sulphate to near saturation. At very high sulphate concentrations the sodium chloride rejection is negative, that is sodium chloride is reconcentrated in the permeate stream and returned to the process, thus improving salt recovery.

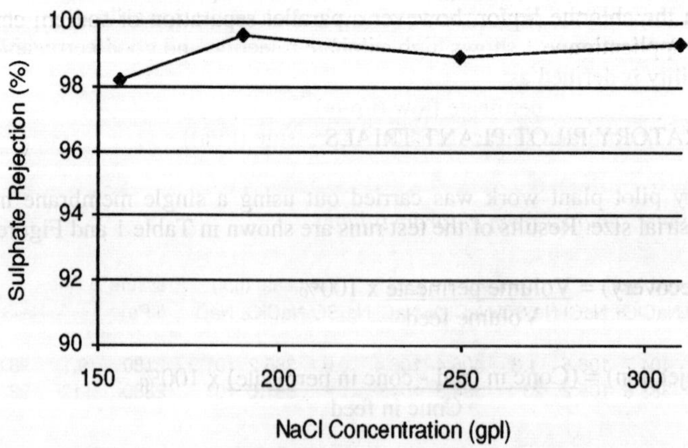

Figure 2 *Sulphate rejection as a function of sodium chloride concentration*

Table 2 *Effect of sulphate concentration on sulphate rejection**

Expt. #	Feed (gpl)		Concentrate (gpl)		Permeate (gpl)		Pressure	% Rec.	% Rej.	
	Na_2SO_4	NaCl	Na_2SO_4	NaCl	Na_2SO_4	NaCl	(kPa)		Na_2SO_4	NaCl
1	1.90	194.50	2.13	192.35	0.00	178.78	2060	10.4	100.0	8.1
2	11.25	189.64	11.88	184.15	0.05	177.93	2040	12.5	99.6	6.2
3	31.10	180.18	34.37	180.76	0.26	173.16	2650	11.4	99.2	3.9
4	47.50	168.50	55.00	164.95	0.50	174.33	3160	10.9	98.9	-3.5
5	71.25	179.00	84.25	178.40	0.88	195.90	3940	11.0	98.8	-9.2

*Single pass experiments, Temperature = 40 °C, pH = 9, 2.5 " X 40 " membrane module, total membrane area 1.7 m^2.

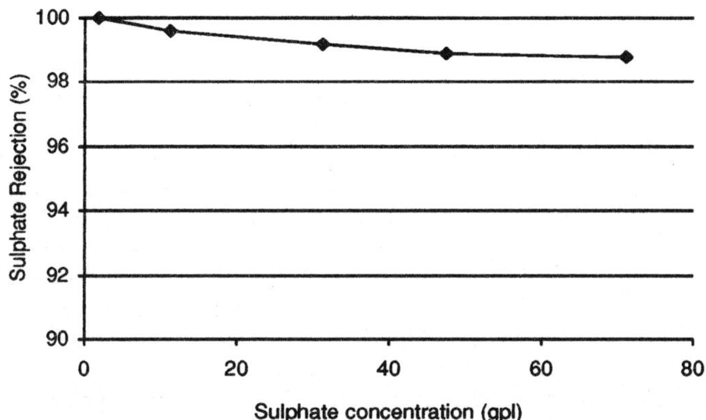

Figure 3 *The effect of sulphate concentration on sulphate rejection*

Table 3 shows the sulphate rejection of a nanofiltration membrane in chlorate liquor. With our nanofiltration membranes sulphate rejection is high and sulphate removal is therefore possible from chlorate liquor solutions.

Table 4 compares the performance of a selection of the nanofiltration membranes that were screened. Membrane 1 shows high sulphate rejection and good permeability.

Permeability is defined as:

$$\frac{\text{permeate flow (l/min)}}{\text{(membrane surface area (m}^2\text{))(pressure (kPa))}}$$

Table 3 *Sulphate removal from chlorate liquor**

Expt. #	Feed (gpl)			Concentrate (gpl)			Permeate (gpl)			Pressure	% Rec.	% Rej.		
	Na_2SO_4	$NaClO_3$	NaCl	Na_2SO_4	$NaClO_3$	NaCl	Na_2SO_4	$NaClO_3$	NaCl	(kPa)		Na_2SO_4	$NaClO_3$	NaCl
1	1.3	404.7	106.5	1.3	406.4	106.4	0.0	385.2	101.9	2180	9.1	98.8	4.8	4.3
2	22.3	383.6	105.2	23.1	384.3	104.0	1.8	381.6	101.7	2880	11.3	92.1	1.9	0.9

*Single pass experiments, Temperature = 40 °C, pH = 9, 2.5 " X 40 " membrane module, total membrane area 1.7 m^2.

Table 4 *Performance comparison for various NF membranes**

Membrane	Feed (gpl)		Concentrate (gpl)		Permeate (gpl)		Pressure	% Rec.	% Rej.	Perm.
	Na_2SO_4	NaCl	Na_2SO_4	NaCl	Na_2SO_4	NaCl	(kPa)		Na_2SO_4	l/(min*m²*kPa)
1	11.25	189.64	11.88	184.15	0.05	177.93	2020	12.5	99.6	0.0048
2	11.81	183.24	12.37	184.19	0.80	179.04	1710	12.0	93.2	0.0057
3	8.38	157.40	8.88	155.90	0.11	140.20	3690	16.3	98.7	0.0033
4	1.52	265.00	1.67	263.00	0.74	253.00	2460	14.0	51.3	0.0043

* 2.5" X 40" membrane modules, for the permeability determination membrane area assumed to be 1.7 m^2 for all membranes.

It is desirable to have good permeability in order to minimize the total amount of membrane required, however, higher permeability is often associated with poor rejection. Membrane 2 has higher permeability than membrane 1 but the sulphate rejection is significantly poorer. Membrane 3 has high sulphate rejection but the permeability is significantly less than membrane 1. Membrane 4 has comparable permeability to membrane 1 but the rejection is very poor.

Table 5 lists some results of our investigations into the removal of other brine impurities by nanofiltration. High rejections are seen for aluminum, sulphate and chromate and somewhat lower rejections are seen for silica and calcium. Mercury in brine at 10 ppm levels was seen to be slightly rejected in some cases and slightly concentrated in others. Practically, this means that mercury cell chloralkali brine, which undergoes nanofiltration for sulphate removal, will not have significantly higher mercury levels in the sulphate rich effluent and may in fact have reduced mercury levels relative to the process brine should the rejection be negative (mercury is preferentially found in the permeate and this is returned to the process).

4 SULPHATE REMOVAL SYSTEM

The Sulphate Removal System is a skid mounted unit with the following equipment: a cooler to provide brine at a temperature below 50°C, a feed tank, a high pressure pump, and filtration modules, which consist of spiral wound filtration elements contained in separate membrane housings. It is designed, primarily, to be used with membrane cell plants where it operates on a side stream of dechlorinated brine of 200 gpl NaCl containing less than 11 gpl Na_2SO_4, see Figure 4. The solution that passes through the membranes contains sodium chloride with trace amounts of sodium sulphate. It is collected in a header and returned to the salt saturator.

Table 5 *Removal of impurities from brine using nanofiltration*

Impurity	Concentration	Rejection	Comments
Aluminum	10 ppm	>85 %	pH 9, NaCl 200-300 gpl
Silica	10 ppm	40 %	pH >7, NaCl 200-300 gpl
Calcium	100 ppm	26 %	pH 2, NaCl 300 gpl
Sulphate	0.5-70 gpl	>97 %	pH 9, NaCl 150-300 gpl
Chromate	5 gpl	87 %	pH 9, NaCl 100 gpl, $NaClO_3$ 400 gpl
Mercury	10 ppm	-40 to 30%	pH 9, NaCl 250 gpl, Na_2SO_4 30-80 gpl

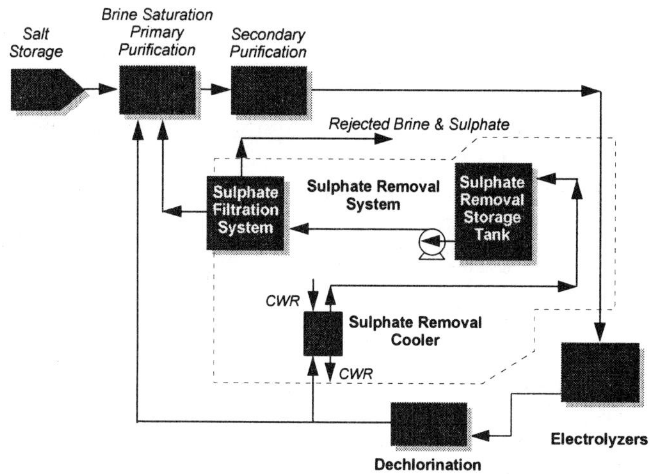

Figure 4 *Sulphate removal within the brine circuit*

The brine is pumped from the spent brine system of a membrane or mercury cell plant, cooled to 45 °C, using a plate and frame heat exchanger and stored in a feed tank. A high pressure pump, with a discharge pressure up to 3700 kPa, transfers the brine to the first membrane module. Brine enters the spiral wound membrane and passes through the membrane spacer channels. The sodium chloride brine and dissolved cations pass easily through the membrane to the low pressure side of the module, where it is collected in an internal pipe and passes out of the membrane housing into a main collection header. This brine, the permeate, is returned to the salt dissolving system.

The sulphate ions in the brine solution are rejected at the membrane surface and pass along the membrane spacer channels, exiting the spiral wound module into the membrane housing and outlet nozzle. The brine exiting the module is the reject brine, which has a lower flow rate than the brine entering the filtration module. This sulphate rich brine enters a second membrane module where more pure brine is removed and the sulphate concentration is further increased in the reject brine.

A number of module stages are connected in series to concentrate the sulphate ions in the solution. The final reject brine concentration approaches the limit of the brine solubility. The sulphate rich effluent is finally discharged as a brine solution containing 200 gpl of NaCl and 80 gpl Na_2SO_4, see Figure 5. The small volume of this effluent (approximately 10% of the original purge stream) will normally be acceptable for most operating plants and precipitation of the sulphate ions or evaporation of the effluent is not normally necessary. However the sulphate ions could be precipitated with calcium

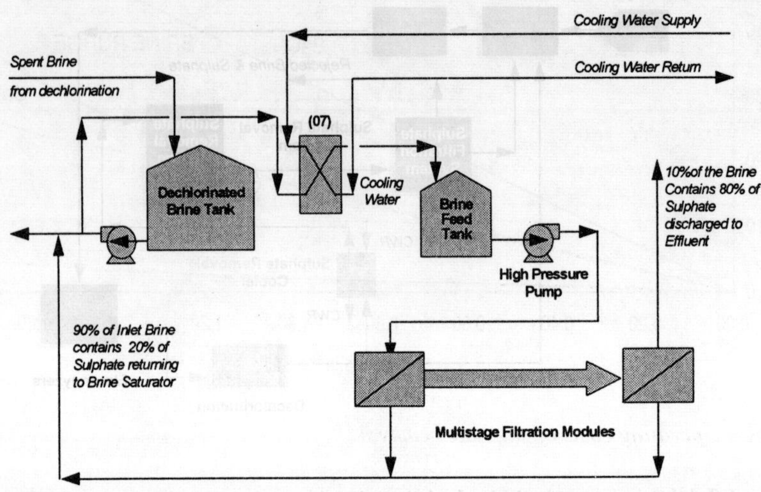

Figure 5 *Sulphate removal process*

chloride or barium chloride, if this is required. Further separation of sodium sulphate from sodium chloride can be achieved by diluting the reject brine and passing this dilute solution through an adjacent filtration system, to produce a higher sodium sulphate concentration and a lower sodium chloride concentration in the effluent stream.

5 SULPHATE REMOVAL SYSTEM COSTS

All sulphate removal systems have two main costs, the initial capital cost and the operating cost.

The capital cost of the Kvaerner Chemetics system will vary for each operating plant. The size and number of filtration modules depends on the operating plant capacity, the concentration of sulphate which dissolves from the raw salt into the brine and the chloralkali process used. The capital cost can be recovered within a short period by the elimination of the chemicals used in sulphate precipitation or the recovery of pure brine from those plants using a purge to limit the sulphate levels in the brine.

As shown in Figure 6, the operating costs of the Kvaerner Chemetics system are low, consisting of electrical power for the high pressure pumps, a small loss of sodium chloride and membrane replacement.

Costs of competing systems are based on Kvaerner Chemetic's plant designs using current chemical costs[6] and a cost for sodium chloride of \$30 per tonne.

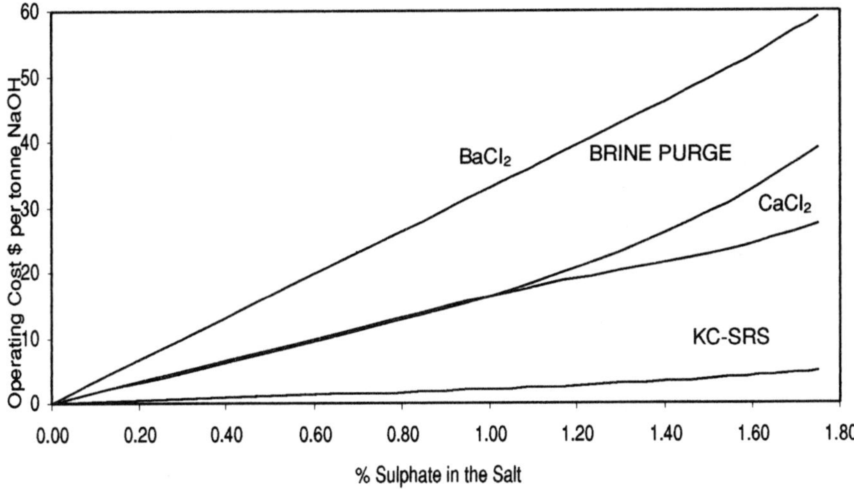

Figure 6 *Operating costs of sulphate removal*

6 SUMMARY

The sulphate removal system described above has the following advantages:

1. A simple skid mounted unit is provided complete with equipment, piping and instrumentation, which makes the system easy to install and connect to an operating plant.
2. The system has a low operating cost.
3. The equipment payback is attractive for most chloralkali plants.
4. There is no chemical addition to the system.
5. The system has simple maintenance procedures and the membranes will not foul with depleted brine.
6. The concentrated sulphate is discharged as a liquid effluent.

REFERENCES

1. U.S. Patent 5,587,083 - assigned to Kvaerner Chemetics Inc.
2. U.S. Patent 5,071,563 - assigned to Kaneka Corp.
3. U.S. Patent 5,093,089 - assigned to Kvaerner Chemetics Inc.
4. R. Rautenbauch and A. Groschi, "Separation potential of Nanofiltration Membranes", *Desalination*, 1990, **77,** 73-84.
5. L.P. Raman, M. Cheryan and N. Rajagopalan, "Consider Nanofiltration for Membrane Separation", *Chemical Engineering Progress*, 1994, **90,** 68-74.
6. Chemical Market Reporter, March 17, 1997.

26
PC CONTROL OF ELECTROCHEMICAL PROCESSES

Ken Woodard, David Cawlfield and Don Loftis

Olin Corporation
650 25th Street
Cleveland, Tennessee 37311

1 INTRODUCTION

Olin Corporation has long been a pioneer in the Chlor-Alkali Industry. On November 25, 1897, the production of chlorine and caustic soda began at the Niagara Falls site of Olin Corporation using Castner Electrolytic Mercury Cell Technology. The rated capacity of the plant using 580 cells was 19.5 tons per day of solid caustic soda and 17.5 tons per day of chlorine. The chlorine was converted entirely to bleaching powder! The plant expanded in 1901 and 1915 adding 1450 cells, increasing the capacity to 90 tons per day of chlorine. Production capacity later expanded to 125 tons per day by increasing the current density up to an average load of 1800 amperes per cell. Liquid chlorine was first produced in 1909, with the first One Ton chlorine container being built for liquid chlorine shipment in 1910.

One hundred years later, in the fourth quarter of 1997, Project SUNBELT, a partnership between Olin Chlor-Alkali Products Division and the Geon Company, will start up 1600 UHDE membrane cells at McIntosh, Alabama, producing 685 tons per day of chlorine. The 1700 I/O devices in the plant will be controlled by four INTEL Pentium PCs, each with a hot back-up shadow, networked together with Ether-net fiber optic communications into a truly distributed control system. OMNX Control software provides the control engine, HMI, and networking capability to make this soft-DCS system a reality.

The rapid development of personal computers (PCs) has brought about a revolution in our workplace and our home. Direct control of processes with the low cost personal computers developed for the masses, is a reality today! How did Olin develop the confidence in PC control to use it for our latest major chlor-alkali investment? Our transparent partnership with a little known (at the time) real time operating system, QNX software, the INTEL processor, and our own design and programming concepts, today known as OMNX Control Software, have allowed a mature fully distributed INTEL Pentium powered software control system to develop.

2 ECHO TECHNOLOGY AND OMNX CONTROL

Olin Technology began using the PC for process control in the early 80s. It began with the utilization of a PC with an 8088 Intel Processor for direct process control during the development of our membrane cell process (ECHO). The requirements were that the process be able to operate unattended most of the time and that the cost of this control had to fit within tight capital cost constraints. Our research said we needed a distributed control system at an order of magnitude lower cost.

The ECHO process produces sodium dithionite solutions electrochemically, using water, sulfur dioxide, caustic soda, and electricity as raw materials. This process was designed to operate on a relatively small scale so that it could be operated near a customer's site. The limited storage life (requiring refrigeration) and dilute concentration (13% by weight) make shipping costs of the product high compared to that of its raw materials. However, to achieve economic operation at a scale of less than 2 MM pounds/year required that the process be able to operate unattended most of the time.

Autonomous control became our new paradigm for design of control systems that operate without constant operator attention. Olin pioneered the first autonomous control system with the commercial demonstration of the ECHO process that began operation in 1987. The OMNX control system was developed to meet its needs for autonomous control. This system provides several features that made autonomous control possible:

- PC-based multi-tasking direct control
- Voice alarms, with interactive telephony
- Alarm monitoring systems that respond to alarms automatically and implement limited artificial intelligence
- Event recording
- High fidelity universal trending
- A multi-user human interface with remote accessibility
- Operators trained to intervene and diagnose problems
- New ways to capture operating expertise

The idea of autonomous control is simple, build a control system that can operate unattended overnight and at weekends. Ideally, this cuts labor cost by about a factor of four. However, the reality is that the system must also recognize when something important happens that needs human intervention, and notify a person (yes, even when they are at home). Even including the extra pay for keeping people on call, and the overtime pay when they occasionally work extra, a factor of three is realistic.

The implementation details of autonomous control are not entirely obvious. Olin was fortunate to have a few individuals responsible for the development of the first autonomous processes who thought that anything was possible with a computer and some programming skills. Based on this experience, they learned the necessities of process control, and invented means to accomplish them.

Autonomous control requires more I/O than traditional automatic control approaches, because this design assumes no manually operated valves or manually read gauges. PC-based control uses non-proprietary high-performance computers to perform control algorithms. Because the PC industry has advanced quickly, microprocessors used in PCs are several generations more advanced than found in leading PLC and DCS control systems. Large scale markets for PCs have pushed the costs of producing them far lower than for the specially designed control processors they can replace.

This highly cost-effective approach is possible only with the use of multi-tasking fault-tolerant operating systems (OS). Olin's OMNX system uses the QNX operating system. The OMNX system allows all control system programming to be changed on-the-fly, so operators can make corrections and improvements without interrupting the control algorithms. Using a multi-tasking operating system, it is possible to schedule the changes so that they occur only during the intervals between control updates. Also it is possible to make control updates take priority over everything else.

The architecture of OMNX control has a process control kernel (pck) that is similar to the OS microkernel in design. It uses Tag names for all devices. The four classes of devices are analog inputs, analog outputs, digital inputs and digital outputs. The process control kernel not only maintains a real-time database of information about these devices, but it actually performs the control functions associated with each.

The analog input device class includes both real analog I/O signals as well as several types of virtual devices, that include totalizers, timers, counters, delays, and mathematical expressions. All analog input devices also share common capabilities, such as alarms, a bad quality indication, and a 32 bit floating point value. Other device classes were designed with a similar structure.

The soft-DCS architecture of OMNX confers several advantages that are essential to control of an electrochemical process.

- On-line changes to any one device are made independently.
- Real and virtual devices are interchangeable (also on-the-fly).
- Complex control system logic can be built incrementally to capture process knowledge.
- New control algorithms, like model-predictive control can be tested along side existing proven methods.

2.1 Voice Alarms and Computer Telephony

Voice alarms are possible and inexpensive because computer vendors have mass produced electronics for PC-based multi-media and voice-mail applications. They are necessary because operators have a new role to play for autonomous systems. Operators intervene when needed, so they are free to perform other tasks the rest of the time. Voice alarms played over a loudspeaker allow the operator to perform equipment checks, simple maintenance jobs, and other activities outside the control room. At night, when the system runs completely unattended, it can call for assistance using dial-out capabilities.

When there is potentially more than one person on call, it is essential to allow the person called to acknowledge that they received the message and will respond. The voice-alarm system we developed not only plays back pre-recorded alarm messages on the phone, but also listens for touch-tone signals indicating acknowledgments or requests for additional information. A remote operator can query the system for information about current alarms so they can respond appropriately. All of this activity is logged.

Telephony can also make use of pagers or cellular telephones, and someone who has acknowledged a call can dial in through a modem to see full details remotely. Voice capabilities are more flexible than mere modem access.

Another application of autonomous control allows one operator to supervise several independent processes. The computer performs autonomous control whenever the control room is empty, eliminating the need for an additional operator to stay in the control room. Voice alarms make communication much easier. In fact, the voice messages can include advice about things that need to be corrected on the plant floor. Another way in which voice alarms can carry more information is by taking advantage of the ability of people to recognize different voices. By recording alarms for different areas using different voices, operators immediately know what area is involved, even before the message has been played. This feature was used extensively in the operation of the liquid gun propellant production.

2.2 Alarm Monitoring

In an attended system, operators often work by dealing with alarms, using different alarms as prompts for when to execute a particular procedure that they know will often correct a problem. Under some circumstances, they may have to shut down the process to avoid hazards to equipment, the environment, or personnel. In an autonomous system, all of these actions must occur automatically. An event-driven alarm monitoring system enables the OMNX control system to accomplish this as well. In this UNIX environment, scripts are written using the shell interpreter, so that any system action can be accomplished easily.

For example, in addition to responding with a series of control actions, a script may wait for particular conditions, automatically start an event logger, send messages to a printer, manipulate files, or anything else that can be accomplished using any of several hundred system utility programs.

When a system operates unattended, it is inevitable that something in the process may happen that requires shutting down the process safely, then calling in someone who can diagnose what went wrong, and fix it. Often though, there is an allowable grace period during which an alarm monitoring system can wait for someone to respond to a call for assistance before taking more drastic action. The alarm monitoring system must be able to interrupt a procedure it has initiated if the grace period has not yet expired. By integrating the alarm monitoring system with voice alarm capability, the person called immediately knows how much time is left.

Another aspect of alarm response is that a human operator is capable of responding to more than one problem simultaneously. The automated alarm monitoring system must always do this, responding to events concurrently, while continuing to monitor new alarms. All of this simultaneous scheduling of alarm responses gets rather complicated, and the system built to handle it exhibits a different sort of artificial intelligence. Not merely applying logic, but rather a real-time intelligence that applies logic to the timing of automated responses.

2.3 Event Recording

When something goes wrong in an unattended system, there is always an opportunity to make improvements. The matter of diagnosing what happened calls for accurate records in which a correct sequence of events is maintained. Event recording must include accurate sequencing of alarm events.

In a direct control multi-tasking system, accurate alarm sequencing and recording is much easier than when using multiple PLCs connected to a central computer. The problem with using PLCs for this purpose is that the control cycles in which alarms are detected and responses occur are asynchronous with communications cycles to the computer. Alarms recorded by the computer in this case may be out of sequence in which they occurred. In a direct control system, alarm sequencing is simple and accurate. The OMNX process control kernel maintains an alarm buffer so that alarm sequences are always maintained, even when alarm-handling functions operate at lower priority than alarm generation. Alarm logging activities are automatically synchronized so that they interleave with the control system updates.

2.4 Learn-Mode Automatic Scripting

"Learn-Mode" describes the macro recording and playback system developed to record operator's actions. To make startups and shutdowns fully autonomous requires a method of capturing the expertise of a human operator first. Unlike a spreadsheet macro language, the timing and logic of an operator's actions must be captured as well, so the output of the system is a script with wait times. Also, the operator must be able to record pending events so this information is also embedded in the script.

Learn mode records every operating change made by an operator, so it can also be used to create operation logs for later review. Yet another use is to reverse-engineer operating procedures by recording what an expert operator does, and use the plain text script file to update process documentation.

2.5 High Fidelity Universal Trending

Data logging capabilities are commonly available in SCADA packages today, but autonomous operations have special added requirements. First, for diagnosis of an event that occurred while no one was watching, it is important that the state of every analog point was recorded. When reviewing records of an unexpected event, important information will be obtained by examining trends that were probably not considered important by the designers. Universal trending requires more memory and storage space, but less foresight than common data logging approaches. If a new point is ever added to the system, it is automatically added to the trend list.

High fidelity is essential for autonomous control. The simple averaging, box car data compression methods used by most historians is not good enough. This type of data compression often misses short-term disturbances that correspond to a significant event. If an operator is not watching, these important events can be ignored. An alternative approach to data compression that is simple to apply is to retain the maximum and minimum values of each point, sampled frequently (about every two seconds) but stored less frequently. When this data is displayed, a signal with a spike will retain its appearance, regardless of the time scale of the view. The result of this style of trending is very much like a pen-style strip-chart recorder. Slowing it down for a noisy process results in fat filled-in tracks on the chart.

Universal trending with min-max data compression meets the minimum requirements for autonomous control. Trend facilities that only retain information while a chart is visible are normally of little value, with one important exception. High speed trending where points are scanned and displayed 5-200 times per second is very useful for tuning fast loops and diagnosing noisy signals. This is a feature that is normally only possible when data acquisition and display are performed by the same computer. With the OMNX system, a specially designed high speed trend server collects data rapidly and can deliver it to display clients across the network. Olin has used this capability to monitor vibration sensors, enabling remote diagnosis of chlorine compressors.

2.6 Multi-User Human Interface, with Remote Accessibility

Remote access is essential for autonomous systems. This access can take place across dial-up lines, local area networks, or wide area networks. With operators and engineers able to respond to events from a distance, autonomous control is safer and more flexible.

The most suitable form of remote access for autonomous control is multi-user access. Multi-user control systems allow two or more persons to share access to control data without affecting one another. While some DOS or Windows based systems can use remote access software to take over control of the machine, this is not the same as multi-user access, where local control is neither lost nor disrupted by remote control.

At first thought it might seem that the need for a highly efficient man-machine interface would be relaxed somewhat for an autonomous system. After all, high fidelity graphics can hardly be cost-effective if there is no one watching them. However, the real requirements for a man-machine interface are much more demanding than we initially expected. Because operators use the system less frequently, ease-of-use and ease-of-learning is more important. Another problem is the overwhelming quantity of information to be displayed.

Our solution to the problems of building a cost-effective MMI for autonomous process control systems builds on the concepts of object oriented programming methods. First each control point is associated with pre-built styles of user interface objects. This approach greatly reduces (or eliminates) the time required to build user interface pages. Another aspect of the object-oriented approach to building the MMI is the concept of multiple concurrent views of the process. Each view provides a different perspective for a subset of the entire process. For example, an alarm summary view focuses attention on either the highest priority or most recent alarms. Once an alarm is selected, controls related to this point are available to the operator.

From the overview, points are grouped into different sections of the process and presented in a tabular form. This form highlights points that are in alarm by color, and provides detailed numerical displays that the operators found most useful. Flowsheet views provide process information in the context of process diagrams. These are most useful to new operators needing training about the basics of the process, but were less useful to experienced operators. In addition to these, trend views, history views, configuration and faceplate views are also available. These tools enabled operators to trace through control logic when necessary to diagnose problems (such as "why is this controller interlocked").

The ability to switch between views in order to focus on different kinds of information is a little like accessing hyperlinks on the Internet. This style of user interaction is a highly effective way of getting to details inside a very complex system.

2.7 Reliability

Reliability of PCs and the I/O hardware has been proven in over 10 years of operation of the four ECHO facilities worldwide. The first installation of Keithley Metrabyte I/O hardware is still operating in the "ECHO" pilot plant facility following more than ten years of service. However, the need for speed resulted in the replacement of the original 286 computer one or two generations of Intel chips ago. The commercial installation of "ECHO" technology occurred in Sao Paulo, Brazil in 1989. Of the four original NEC 386/20 computers, three continue in service today, following eight years of service. One computer was replaced by a locally manufactured unit in order to gain experience with up-to-date technology, so when the need develops, local supply options can be exercised. The original operating system and Keithley/Metrabyte I/O hardware are still in service. All technical support has been conducted remotely from the OMNX headquarters in Charleston, TN.

The second "ECHO" installation started up in Charleston, TN in 1990. Of the four original NEC 386/25 computers, one remains in operation today. One of the units experienced a failure of a component that required replacement. This installation used OPTO-22 Pamux I/O components. Service life of these components has been excellent.

3 HAN PRODUCTION FACILITY

About the same time, 1989, PC control was installed with the U.S. Army's hydroxylammonium nitrate production facility at Olin's Charleston, Tennessee site. The installation of two NEC 486/25 machines has provided exceptional process control in the manufacturing of a liquid gun propellant. OMNX software controls a complex membrane and mercury cathode electrolytic process that produces the hydroxylammonium nitrate (HAN) from nitric acid, water, and electrons. This system required somewhat more complex control than the "ECHO" technology. The use of on-line process analyzers, integrated with conductivity and pH sensors, resulted in a total of about 1000 I/O points, most of them analog inputs. Opto-22 Pamux hardware was chosen for the installation due to the flexibility of on-line maintenance of individual input and output modules. Hydroxylammonium nitrate solutions can decompose rapidly and energetically if the process parameters are not carefully controlled. This process certainly requires a high degree of reliability. Extremely accurate control of excess acidity during the manufacturing process is essential to obtain optimum process yields. Critical control includes the excess nitric acid in the catholyte of the electrolytic cell, control of the excess acidity following ion exchange neutralization columns, and maintaining meticulous temperature control within the manufacturing environment. The mission critical nature of the process control required the development and installation of a hot standby redundant shadowing computer system that could take control in the event of hardware failure. The system was challenged several times during the operation of the facility, and each time the process was controlled.

The HAN process has three maufacturing units, electrolysis, neutralization, and concentration. The product from the electrolysis unit contains about 1% excess nitric acid that must be removed prior to concentration. A continuous neutralization step consisting of a series of ion exchange units produce hydroxylamine, the free base, that is used to provide control of the excess acidity, in the range of 0.01 wt % prior to concentration. A two-stage pH control system using operator calibration and reset was required to achieve the minus 0 plus .01 control required for product stability.

Another aspect of the HAN manufacturing was the very high degree of complexity of interlock strategies used to assure the safe operation of each part of the facility. This complex layer of protection, combined with full on-line documentation prevented operators who used the system infrequently from making critical mistakes. Internally, approximately 1000 additional virtual I/O points and over 3000 alarms were configured to implement elements of this control strategy. The HAN process benefited from autonomous control capability by allowing one operator to handle this large and complex process by himself including analytical duties that required him to be away from the control room a high percentage of the time.

The facility provides reliable production of liquid gun propellant to exacting quality specifications. The total production facility is operated by one operator per shift, including the use of an ICP-MS for in-process quality monitoring. The process control was also used to establish process safety envelope parameters that provide PSM guidelines for compliance with OSHA 1910.119. OMNX software is a mechanism that provides comprehensive, instantaneous information, limiting your liability while keeping you inside safety and production windows.

4 SODIUM CHLORATE PRODUCTION

Huron Tech Corp. chose OMNX control for their first phase of their sodium chlorate plant in 1995, producing 40,000 tons per year. The plant was constructed in less than six months and operated on one PC. The second and third phases of expansion of the plant that tripled production also used OMNX control with individual PCs for each production plant. Today, the PCs are networked together into a distributed system for all three process units.

5 SUNBELT CHLOR-ALKALI PRODUCTION FACILITY

The first phase of the SUNBELT project started up in January of this year with the commissioning of the Sodium Sulfate Crystallization Process. The process was designed and constructed by Wheelabrator HPD and controlled by OMNX Control System. The architecture of the OMNX control product allows addition of new control nodes without interruption of the existing operations. The software can be modified on-line to include new points or change existing control configurations or strategies, making it ideal for phased commissioning of the facility.

6 THE FUTURE

Today's PC has enormous computing power. Coupled with highly reliable operating systems and mature control software provided by OMNX Control, the PC provides a platform to utilize non-proprietary I/O devices to construct low-cost, highly efficient control systems. One CPU can service up to 9192 I/O per computer, with access times as low as 1 millisecond, that can include up to 256 controlling computers on a distributed network. The era of the Soft-DCS has only begun to tap the resource capability of today's microcomputers.

Consolidation of the MES and process control systems into one system, called Open Control Systems (OCS) is the next step in the evolution of PC based control.

OMNX OCS software, with its seamless data exchange, networking and data collection done with peer-to-peer process control provides the platform for this evolution to begin immediately. Enterprise control and connectivity between control platforms will become a reality as open control systems mature in the next century.

Index of Contributors

Subject Index